Microbiology and the Spontaneous Generation Debate During the 1870's

by
William Glenn Vandervliet

Coronado Press   1971

Copyright © 1971
by
Glenn Vandervliet

Standard Book Number 87291--020--2

Published by
Coronado Press
Box 3232
Lawrence, Kansas 66044
—All rights reserved—

Manufactured in the USA

Acknowledgments

Since 1964 when I took Dr. Robert C. Stauffer's course in Modern Science at the University of Wisconsin, I have been interested in the spontaneous-generation controversy which has been examined by historians through the 1860's. However, the controversy over the spontaneous generation of microbes from non-living, organic matter continued throughout the 1870's. This study has as its purpose the examination of the debate of that decade and its influence upon the development of microbiology.

Many people have read this essay in its various stages of preparation. Special thanks should go to O.N. Allen, Nikolaus Mani, Glenn S. Pound, and Robert C. Stauffer, all of the University of Wisconsin. Their encouragement, guidance, and helpful criticism were invaluable in writing this work, which was originally my doctoral dissertation in the history of science.

In the preparation of this work for publication I wish to thank H. Lewis McKinney and John E. Longhurst of the University of Kansas for their kind assistance. My wife, Kay Elaine Vandervliet, has been a loyal helpmate and faithful typist throughout. Any shortcomings herein are, of course, my own responsibility.

Finally, I am indebted to the U.S. Public Health Service for Fellowships 1-Fl-GM-29, 050-01 and 5-Fl-GM-29, 050-02 for support in this research.

GV
1971

Table of Contents

I. Introduction 13
II. Microbiology in the Early 1870's 15
    (1) Technical Limitations 16
    (2) Gaps in Microbial Knowledge 18
    (3) Ineffective Heat-Sterilization Procedures 20
III. The Turnip-Cheese Episode 23
IV. The Hay-Infusion Episode 43
V. The Urine Episode 55
VI. Factors Determining the Thermal Resistance
    of Infusions 65
    (1) Circumstances of the Sterilization Process 65
    (2) Characteristics of the Media 68
    (3) Role of Different Kinds of Microbes 72
VII. Heat-Sterilization Procedures of the 1870's 77
    (1) Prolonged Boiling at $100°C$. 78
    (2) Heat-Treatment above $100°C$. 79
    (3) Discontinuous Heating or Tyndallization 81
    (4) Dry-Oven Method of Heat Sterilization 85
Conclusion 89
Footnotes to Text 92
Bibliography 107
Index 129

Chronology

1870 (September 13) T.H. Huxley's Liverpool Address on "Biogenesis & Abiogenesis."
1871 H.C. Bastian publishes *Modes of Origin of the Lowest Organisms* & coins term "archebiosis" for heterogeny.
1872 Foundations of bacteriology laid by Ferdinand Cohn's monograph, *Bacteria: The Smallest of Living Organisms.*
1873 (January) Burdon-Sanderson certifies Bastian's claims that living microbes can be found in organic solutions which have been heated to 100° C. for 5 to 10 minutes.
1873 (May through summer) Cohn begins to study cheese manufacture and butyric fermentation. He discovers bacterial endospore formation in *Clostridium butyricum.*
1874 Roberts' "Studies on Biogenesis" interest Cohn and Tyndall.

1875 (autumn) Bastian begins work on alkaline urine and heterogenesis.
        Tyndall experiments on purely liquid infusions.
        Cohn begins study of hay infusions.
1876 (January) Tyndall insists that "five minutes boiling" is a reliable method of heat sterilization.
1876 (February 10) Bastian calls attention to discrepancy between results presented by Burdon-Sanderson and by Tyndall.
1876 (April 30-May 3) Koch visits Cohn at Breslau to demonstrate life history of anthrax pathogen, *Bacillus anthracis.*
1876 (July) Cohn publishes his July, 1876, *Beiträge zur Biologie der Pflanzen (Contributions to the Biology of Plants)* containing Cohn's paper on *Bacillus subtilis* and Koch's paper on *Bacillus anthracis.*

1876 (late September) Social meeting between Cohn and Tyndall. Perhaps this was the occasion when Tyndall received his copy of Cohn's journal.

1876 (mid-October) Tyndall's renewed study of organic infusions reflects Cohn's influence. Tyndall soon begins to question efficacy of 5 minutes boiling as a dependable sterilization procedure.

1876 (October 19) In his Glasgow Address Tyndall refers to Koch's work.

1876 - 1877 (winter) After bringing dried hay into his laboratory at the Royal Institution, Tyndall encounters difficulties with his infusion experiments.

1877 (January) Tyndall left Royal Institution to go to Jodrell Laboratory at Kew Gardens. There he repeats his infusion experiments obtaining his original results.

1877 (January 18) Tyndall finally admits that 5 minutes boiling is not a reliable method of heat sterilization.

1877 (February 14) Tyndall describes his technique of discontinuous heating in a letter to T.H. Huxley, Secretary of the Royal Society of London.

1877 (May) Tyndall publishes report on hay infusion.

1877 (July) Bastian and Pasteur meet in Paris for proposed contest over neutralized urine and alleged spontaneous generation.

1878 Using the dilution method, Lister isolates first truly pure culture of a bacterium. The microorganism is *Bacterium (Streptococcus) lactis*, a milk-souring microbe.

Chapter 1
Introduction

During the 1870's microbial investigators prepared organic solutions from a wide variety of plant and animal substances to be used as liquid culture media. Some of these fluids — notably turnip-cheese infusion, hay infusion, and urine — were unique in that they frequently possessed unusual resistance to ordinary heat-sterilization techniques. In some instances microorganisms actually appeared within a few days after flasks containing these infusions had been boiled at 100°C. for five to ten minutes and then hermetically sealed toward the end of ebullition.

Thomas Henry Huxley (1825-1895) directed his attention to these puzzling cases of microbial thermal resistance in his Presidential Address at the Liverpool Meeting of the British Association for the Advancement of Science. In those remarks delivered on 13 September 1870, Huxley discussed *Biogenesis* and *Abiogenesis*, which he defined respectively as (1) "the hypothesis that living matter always arises by the agency of preexisting living matter" and as (2) "the contrary doctrine — that living matter may be produced by not living matter."[1] Huxley concentrated on claims that organic solutions in hermetically-sealed vessels subjected to "great and long-continued heat, have sometimes exhibited living forms of low organization when they have been opened." To account for these alleged occurrences of microbial spontaneous generation or heterogeny, Huxley suggested that "there must be some error about these experiments" since "they are performed on an enormous scale every day with quite contrary results" in the food-preservation industry. Upholding the germ theory, he cautioned that microorganisms possess variable heat resistance and that such resistance is influenced by physical and chemical characteristics of the culture medium.[2]

Huxley's comments on microbial spontaneous generation marked the beginning of a decade-long examination of both sides of the controversy over the *de novo* origin of microbial life from non-living, organic matter. In the subsection meetings which followed Huxley's speech, the spontaneous-generation controversy was clearly the main attraction.[3]

This study will discuss the men and events which led to the discovery of bacterial endospore formation, to the recognition of the many factors which influence microbial thermal resistance, and to the development of efficient and effective means of heat sterilization. The scope of this work will be confined to three significant episodes during the 1870's and to the most important investigators who played some role in the turnip-cheese-infusion, hay-infusion, and urine episodes. The figures who participated in one or more of these three episodes were: Henry Charlton Bastian, John Scott Burdon-Sanderson, Ferdinand Julius Cohn, Robert Koch, Edwin Ray Lankester, Louis Pasteur, William Roberts, and John Tyndall. Of the important scientists in our story four were trained as physicians: Bastian, Burdon-Sanderson, Koch, and Roberts. Pasteur was a chemist turned microbiologist. Tyndall was a physicist. Lankester attained scientific eminence as a zoologist and comparative anatomist. Only Cohn was actually pursuing microbial researches as a logical outcome of his training as a botanist. Five of these investigators were Englishmen while Robert Koch and Ferdinand Cohn were Germans and Louis Pasteur was, of course, a Frenchman.

Chapter 2
Microbiology in the Early 1870's

One of the major challenges confronting microbial researchers in the early 1870's was the very nature of the subjects they were attempting to study. Minute size, wide distribution, difficulty of isolation, microbial competition, morphological questions, and confusion over the life histories of these organisms provided difficulties for these investigators. A representative contemporary discussion of the hurdles facing microbial workers is found in Wyville Thomson's Presidential Address delivered before the Botanical Society of Edinburgh on 14 November 1872. Thomson stated that "these plants are extremely minute, and their investigation requires great skill in manipulation, and great practice." He observed that "they are enormously abundant, and their multiplying germs of all kinds are so minute and all-pervading that it requires the utmost experimental dexterity to separate them, to sow them, and still more to exclude them." Commenting upon the difficulty of obtaining pure cultures, Thomson remarked that "if we attempt to select and sow one species, ten to one the seed is mixed with the seeds of a multitude of weeds, and if during the process we allow the most indirect and instantaneous communication with the open air, instantly the enemy sows tares among our wheat, and one of these, probably more vigorous than the others, in the course of an hour has cut short its weak struggle for life." Thomson further cautioned that "the form of these plants requires very careful study — some parts of them, such as the universally diffused mycelium, are undistinguishable in different species; and so are the gemmules, conidia, and spores examined singly." Referring to classification problems, he asserted that "it is often only when the entire 'fructification' is present that distinguishing characteristics exist which one can grasp." Another complication was that "most of these plants present some form of the singular phenomenon of pleomorphy; perhaps not more so than other plants, but slight differences in form tell greatly in such simple and critical organisms." Thomson concluded that "it is therefore not the appearance of the particular mould-fungus at any one time which we have to consider, but its life history" because these minute plants "present different forms at different periods of growth, and under slightly different circumstances."[1]

Thomson's remarks characterize the difficulties challenging investigators who explored the biology of very small fungi. Even more severe were the problems faced by workers engaged in the new science of bacteriology in the early years of the 1870's.

Technical Limitations

Technical problems impeded the progress of microbiology during this period. Microscopes were particularly inadequate, and in 1872 Ferdinand Cohn complained that the study of bacteria was similar to the plight of the lost traveler wandering at dusk through unfamiliar territory. He observed that "the strongest of our magnifying lenses, the immersion system of Hartnack, gives a magnifying power of from 3000 to 4000 diameters" but that "even under this colossal amplification the smallest bacteria do not appear larger than the points and commas of good print." "Of their internal parts little or nothing is to be distinguished," Cohn lamented, "and even their existence would for the most part remain hidden, did they not live in such gregarious masses."[2]

Two years later William Roberts expressed similar frustrations regarding the limitations of the microscope and the difficulty which he had experienced in attempting to study microbial germs. He remarked that "the objects sought after are so minute and so wanting in characteristic forms, that such a search, with our present instruments, appears well-nigh hopeless."[3]

Relying on the vegetable dye, indigo, or on carmine derived from cochineal insects, inadequate staining procedures likewise retarded microbial research. By 1874, however, Carl Weigert (1845-1904) was using aniline dyes in the bacteriological laboratory. The first aniline dye was mauve or aniline purple and was produced in 1856 by William Henry Perkin (1838-1907). Other similar products soon followed. By the beginning of the 1870's basic fuchsin, safranin, methyl violet, aniline blue, eosin, and methyl green were available as microbial stains.[4] Weigert's efforts received great impetus through the work of Robert Koch at the close of the decade.[5]

The lack of pure-culture techniques formed a third technical impediment. Indeed, the first truly pure culture of bacteria is believed to have been obtained by Joseph Lister (1827-1912) in 1878 when he employed the dilution method in order to isolate *Bacterium (Streptococcus) lactis*, a milk-souring microbe.[6] Solid nutrient media and practicable pure-culture techniques were not introduced until 1882 through the efforts of Robert Koch. Later improvements arose from suggestions made by some of Koch's students.

Sir William Watson Cheyne summarized the technical handicaps to microbial research in 1876 at the Edinburgh Infirmary. It depended on liquid culture media and was hampered by lack of staining techniques, oil-immersion lenses, and proper incubators. Initially using milk as his cultivating fluid, Cheyne soon switched to meat infusion or cucumber infusion. He also used boiled flasks capped with cotton-wool plugs sterilized by treatment with ether and heated carbolic acid.[7]

Chapter 2

Gaps in Microbial Knowledge

Gaps in the available knowledge of the biology of microorganisms provided further stumbling blocks to microbial investigators. The concept of bacterial "germinal matter" or of bacterial "germs" was very hazy until well into the decade of the 1870's. This lack of precision in the minds of early bacteriologists caused confusion between adult microorganisms and spores, from which at least the higher fungi were then known sometimes to arise.

As early as 1862 Louis Pasteur had described "corpuscules organisés" distributed in the air. Pasteur found that these minute, spherical particles had diameters in the range of 1/100th of a millimeter, and he had to conclude that his corpuscles probably were germs of a number of different species of microbes. He was unwilling, however, to designate any particular corpuscle as being a definite spore or egg of a given microbe.[8]

Writing in 1877, John Burdon-Sanderson pointed out that Pasteur's "corpuscules organisés" were "finished organisms" and were therefore adult microbes. These were to be distinguished from the bacterial germs of John Tyndall which were the endospores giving rise for example to the vegetative cells of *Bacillus subtilis*.[9]

Man's ignorance of the exact nature of microbial germs was further revealed in an article written in April, 1874, by William Roberts. He remarked that "the word *germ* is used simply in the general sense of a particle endowed with the power of provoking germination in a suitable medium."[10] In this paper Roberts explained in detail his concept of microbial germs as of 1874, describing germs as being susceptible to heat destruction and hence organic in nature. In addition, germs were subject to mechanical filtering and to the laws of gravity; therefore they were solid particles. He also suggested that atmospheric germs consist of true spores as well as the microbes themselves. Like Pasteur, however,

Roberts could not identify individual atmospheric particles with particular microorganisms.

> The exact nature of these germinal particles is, however, not a matter of actual knowledge. It is evident that they are organic, because of their easy destructibility by heat. It is further evident that the atmospheric germs are solid particles, for they can be mechanically filtered from the air, and they are incapable of ascending against gravity. It may be assumed that they consist partly of true spores and partly of the organisms themselves, floating amid the dust of the atmosphere or mingled with the molecular matter always present in ordinary water. But it cannot be said that they have ever been actually seen and identified. The ingenious attempts of Pasteur and others to demonstrate germs in the air are manifestly illusory. Like them I have repeatedly collected air-dust and found abundance of molecules, circles, spheres, and particles of various kinds under the microscope; but these could not be identified as true spores, nor distinguished from particles of inert dust.[11]

Besides the confusion which existed in the early 1870's over the distinction between the adult bacterial cell and the "germs" from which many microbial workers contended that some adult bacteria arose, there were other aspects of the biology of microbial life which were not well understood.

Microbial classification, especially bacterial classification, was in a most tentative state. Ferdinand Cohn initially attempted to classify the bacteria on morphological grounds, but he soon emphasized the necessity of using physiological characteristics as well in his classification efforts.[12]

The terminology used to describe the various kinds of microorganisms was extremely imprecise until well into the 1870's. For example, Louis Pasteur's intense interest in the practical applications of his microbial investigations was not always accompanied by great precision in microbial terminology and classification.[13] The descriptions provided by Pasteur of the microbes which he studied were a great source of vexation to Ferdinand Cohn. Cohn, a more classification-minded microbiologist, constantly strove for extreme accuracy in the naming and in the description of bacteria and other microbial forms.[14]

The specificity of microbes was also in a state of confusion. For a time, even Joseph Lister accepted the concept of pleomorphism. In this view one kind of microbial life was thought to undergo transition into another kind of microorganism. On 7 April 1873 Lister addressed the Royal Society of Edinburgh and referred to his conversion to the position that bacteria might arise from filamentous fungi.[15] By the second half of the 1870's, Lister readily admitted that he had been mistaken in supporting the apparent phenomenon of microbial pleomorphism; but his error of observation can be readily understood. Pure cultures of microorganisms and especially of bacteria were rare before the early 1880's. Consequently, it was easy to consider two adjacent, but distinctly different kinds of microorganisms, to represent transitional forms of one microbe giving rise to another kind of microorganism.[16]

Nutritional as well as oxygen requirements of microbial life were little understood early in the 1870's. There was also a need for detailed study of the distribution of microorganisms in water and in the atmosphere. Until these relationships could be firmly established, controversy would continue over the distinction between bacterial cells in the fully-developed condition and bacterial germs, from which the germ theorist contended that adult bacteria sometimes arose.

Ineffective Heat-Sterilization Procedures

Microbial research in the early 1870's frequently relied upon the use of what were subsequently learned to be ineffective methods of heat sterilization. In the year 1870 there was general agreement that 60°C. or 70°C. were sufficient temperatures to kill microbial life including adult bacteria. Furthermore, almost everyone agreed that boiling a solution of plant or animal substance (that is, an infusion) for five or ten minutes generally rendered the liquid organic material free of microbial life including adult bacteria. For the most part, the food-preservation industry operated with great success by simply boiling and then hermetically sealing food containers toward the end of ebullition.[17]

Certain kinds of organic solutions, however, could not with certainty be rendered free of bacterial life by the widely employed heat-sterilization procedure of boiling for five or ten minutes followed by hermetic sealing toward the end of ebullition. Of the many organic solutions used as liquid culture media by microbial investigators in the 1870's, turnip-cheese infusion, hay infusion, and urine presented especially puzzling instances of thermal resistance. They were vexing but revealing episodes in the history of microbiology.

Chapter 3
The Turnip-Cheese Episode

Studies made by Henry Charlton Bastian, John Burdon-Sanderson, Edwin Ray Lankester, John Tyndall, and Ferdinand Cohn on turnip-cheese infusion during the early 1870's demonstrated the need for detailed study of individual microorganisms and their behavior under adverse environmental conditions. Turnip-cheese infusion was one kind of liquid culture medium, prepared from plant and animal materials, which possessed troublesome properties for microbial investigators. As is commonly known today, cheese is a product of microbial activity, and several microorganisms are involved in cheese manufacture.[1] Ferdinand Cohn discovered that the addition of cheese particles to turnip infusion introduced extremely heat-resistant bacteria into that infusion. His life-history studies of specific microorganisms clarified the empirical investigations conducted by British scientists on turnip-cheese infusion.

In the year following Huxley's Liverpool Address a little book was published which offered the term *archebiosis* for the process by which microorganisms "may arise *de novo* in certain fluids containing organic matter, independently of preexisting living things."[2] The author of the book was the London physician Henry Charlton Bastian (1837-1915). One of the first English neurologists, by 1871 when he was just thirty-four years of age Bastian had acquired an impressive number of titles: he was Assistant Physician to University College Hospital, Assistant Physician to the National Hospital of the Epileptic and Paralysed, Fellow of the Royal College of Physicians, and Professor of Pathological Anatomy at University College, London. Bastian contributed numerous articles to the journal *Nature*, and he also wrote several books with detailed accounts of experiments.[3] He regarded these experiments as proof of the existence of bacteria in non-living, *organic* solutions subsequent to the treatment of such solutions by methods generally considered to destroy all previously-existing microbial life. Bastian's abundant enthusiasm and his desire to establish his cherished belief in the *de novo* origin of microbial life from non-living materials led to some unfortunate results. For example, he developed an annoying tendency to present the factual findings of other investigators under the cloak of his own interpretations on the origin of microbes. This characteristic eventually made Bastian the international opponent of those men who were attempting to place the new science of microbiology upon a sound scientific foundation.[4] Somewhat pathetically, Bastian continued to write voluminously in support of the spontaneous generation of microorganisms through the first decade of this century.

Early in the 1870's Bastian claimed that his turnip-cheese infusion, even after being boiled for ten minutes and then hermetically sealed during ebullition, gave rise to microorganisms within a few days. He maintained that adult bacteria were killed by exposure to $60°C$. Thus he reasoned that the temperature of boiling water should be more than sufficient to sterilize turnip-cheese infusion.[5] Not convinced by the analogy drawn from the heat-resistant spores of the higher fungi,[6] Bastian demanded that the advocates of the germ theory provide direct proof of the durable attributes which they assigned to bacterial germinal matter.[7] Since such bacterial germinal matter was invisible under the 500- to 1200-diameter magnifying power of microscopes then used in England, Bastian's reluctance to accept all which the germ theorists would attribute to their bacterial "germs" was to some extent justified.[8]

Even before 1870, British periodicals contained references to the extreme heat resistance of some microbes. For example, in 1868 Edward Frankland wrote that "boiling even for several hours cannot be relied upon for the destruction of such germs, some of which have recently been shown to retain their vitality after four hours boiling; in fact there can now no longer be any doubt that, as contended by M. Pasteur, the cases of so-called spontaneous generation have all had their origin in ignorance of the excessive tenacity of life in the germs of the lowest organisms."[9] Despite such statements, Bastian was not alone in his position that 100° C. should be lethal to bacterial germinal matter. Huxley, an advocate of the germ theory, likewise assumed that 100° C. would kill all forms of microbial life. In fact, at a Sectional Meeting following the delivery of his British Association Address at Liverpool in 1870, Huxley had stated: "I cannot be certain about other persons, but I am of opinion that observers who have supposed they have found Bacteria surviving after boiling have made the mistake which I should have done at one time, and, in fact, have confused the Brownian movements with *true living* movements."[10] Huxley's attempt to account for Bastian's claims on the grounds of mistaken observation or of errors of technique was to be a frequent approach employed by Bastian's opponents.

Bastian's work on turnip-cheese infusion received its first unbiased review in December, 1872, by the British physiologist John Scott Burdon-Sanderson (1828-1905), one of Bastian's colleagues at University College, London.[11] Burdon-Sanderson had been trained in medicine at the University of Edinburgh and in physiology at Paris where he studied under Claude Bernard. In 1871 he was appointed as the first Director of the Brown Institution, which was a division of the University of London responsible for conducting research in animal pathology and in physiology. Three years later Burdon-Sanderson accepted the Jodrell Professorship of Physiology at University College, London. After his London positions he went to Oxford in 1882 where he served as the first Waynflete Professor of Physiology. Thirteen years later Burdon-Sanderson became Regius Professor of Medicine at that institution.[12]

Burdon-Sanderson had watched a friend of his prepare, boil, and hermetically seal various organic solutions all of which Burdon-Sanderson had found to remain free of microbial life. He then sent samples of these solutions to Bastian who replied that since the infusions differed from his own they had not been prepared satisfactorily. Bastian offered to perform his own infusion experiments under the supervision of Burdon-Sanderson. The latter expressed a willingness to give Bastian a hearing and to publish an account of the results which Bastian obtained. However, Burdon-Sanderson was adamant concerning his desire to remain uninvolved in the dispute raging between Bastian on the one hand and the advocates of the germ theory, such as John Tyndall, on the other hand: "I have hitherto taken no part in the controversy relating to spontaneous generation, and do not intend to take any."[13]

The article published by Burdon-Sanderson in the 9 January 1873 issue of *Nature* is noteworthy for several reasons. Burdon-Sanderson was careful to describe the conditions of Bastian's turnip-cheese experiment with great precision, although as Lankester later observed, Burdon-Sanderson could have been more specific about the quality, quantity, and methods of handling the cheese employed by Bastian.[14] Burdon-Sanderson showed great initiative in suggesting that in his third set of experiments Bastian use flasks previously heated and sealed at 250°C. in order to exclude the possibility of contaminants being present on the walls of the vessels prior to filling them with the infusion. Burdon-Sanderson reaffirmed his neutrality regarding heterogenesis. He also clearly certified the exactitude of Bastian's methods of operation and the accuracy of the results which he obtained. Finally, he described in detail the microorganisms present in Bastian's treated flasks.

Burdon-Sanderson observed that in the first series of experiments carried out on 14 December 1872 Bastian had used all parts of the turnips including the rind and had prepared an infusion with a specific gravity of 1012. The turnip infusion was acidic, and Bastian added liquor potassae (potassium carbonate) to one half of the infusion in order to neutralize it. He then filled four flasks with the turnip infusion, two flasks being acidic in reaction, two flasks being neutral. To one flask of each group he added about two grains of pounded Cheddar cheese. All four flasks were boiled over a Bunsen burner for five minutes and then hermetically sealed with a blowpipe flame. Of these flasks of Bastian's first series of experiments Burdon-Sanderson found evidence of microbial life only in the flask of turnip-cheese infusion which had been neutralized with liquor potassae prior to boiling. Examining this flask three days after its preparation and sealing, Burdon-Sanderson wrote that: "the liquid was crowded with moderately sized Bacteria, which exhibited active progressive movements. There were also Leptothrix filaments [i.e. long threads of bacterial cells]."[15]

Because Bastian had anticipated changes in all four turnip flasks, Burdon-Sanderson watched him carry out a second series of investigations on 20 December 1872 in which all the turnips used by Bastian were peeled. Hence, this batch of turnip infusion contained no rind, but was composed entirely of filtrate from boiled slices of the central portions of the turnips. In this case all the flasks showed evidence of microbial life after five minutes boiling and sealing. Burdon-Sanderson's description of the bacteria which he observed in this series of investigations is extremely important because the accuracy of his report was to be seriously questioned by John Tyndall until 1876.[16]

In the third series of investigations carried out by Bastian on 27 December 1872 Burdon-Sanderson provided him with two flasks previously heated and sealed at 250°C. Bastian broke the points of these vessels under the surface of a neutral turnip-cheese infusion, filled the calcined flasks, boiled one for five minutes and the second for ten minutes and sealed them both during ebullition. Both of these flasks Burdon-Sanderson described as "full of Bacteria, whilst Leptothrix existed in abundance in portions of the scum, together with granules of various sizes which refracted light strongly."[17]

Burdon-Sanderson's descriptions of the microorganisms which survived Bastian's five or ten minutes boiling and sealing treatment are very instructive. From his comments we know that the adult bacteria were of moderate size, that they formed Leptothrix threads (i.e. strings of rod-like bacteria attached end to end), and that they were motile organisms. Furthermore, Sanderson's reference to "granules of various sizes which refracted light strongly" is most interesting as we shall see.

While refusing to comment on Bastian's turnip-cheese experiments "as to their bearing on the question of heterogenesis," Burdon-Sanderson did state the exact significance which Bastian's demonstration held for him. "The accuracy of Dr. Bastian's statements of fact, with reference to the particular experiments now under consideration, has been publicly questioned," he wrote, "and I myself doubted it, and expressed my doubts, if not publicly, at least in conversation." In conclusion, he wrote:

> I am content to have established — at all events to my own satisfaction — that, by following Dr. Bastian's directions, infusions can be prepared which are not deprived, by an ebullition of from five to ten minutes, of the faculty of undergoing those chemical changes which are characterized by the presence of swarms of Bacteria, and that the development of these organisms can proceed with the greatest activity in hermetically-sealed glass vessels, from which almost the whole of the air has been expelled by boiling.[18]

One of the first investigators to respond to Burdon-Sanderson's January, 1873, report on Bastian's turnip-cheese work was Edwin Ray Lankester (1847-1929). Upon graduation from Christ Church, Oxford, Lankester travelled as a Radcliffe Fellow, displayed an interest in marine biology at Naples, and in 1872 became a Fellow and Tutor at Exeter College, Oxford. While at Oxford, Lankester worked on turnip-cheese infusions with Dr. C. C. Pode, Demonstrator to the Regius Professor of Medicine. By 1874 Lankester was appointed Jodrell Professor of Zoology at University College, London, and the following year he became a Fellow of the Royal Society of London. Lankester distinguished himself as a zoologist, leaving his London post in 1891 to become Linacre Professor of Comparative Anatomy at Oxford. He also was editor of the *Quarterly Journal of Microscopical Science* from 1878-1920.

Shortly before the publication of Burdon-Sanderson's article, Lankester and Pode had carried out experiments on turnip infusion and had obtained results contradictory to those of Bastian. Lankester was determined to reconcile the discrepancy between his results and those of Bastian. He began by publishing in *Nature* a request for further information concerning Bastian's exact procedures:

> It seems important that it should be known (1) what kind of cheese was used, (2) about how much to each fluid ounce of turnip infusion, (3) to what extent the cheese was "pounded" before addition, and whether particles of cheese visible to the naked eye, and of what approximate size, were present in the infusion during its boiling? 4) whether the turnip solution was strained before the addition of the cheese, and whether it contained obvious solid particles, and of what size.

A very interesting aspect of Lankester's query for specific instructions on the preparation of turnip-cheese infusion was his closing paragraph.

> It is desirable to call to mind that Pasteur himself and others have recorded experiments regarded by them as demonstrating the survival of the Butyric form of Bacterium or its germs, after exposure to temperatures of 100° or even 105°C.[19]

Lankester's comment concerning the Butyric organism is interesting in that it seemed to play little or no role in Lankester's approach to the turnip-cheese episode. Quite the reverse was to be the case with the investigations of Ferdinand Cohn in Germany which we shall discuss later.

Throughout the year 1873 many British investigators studied the behavior of organic infusions in response to high temperatures. For example, from 1872 to 1874, Dr. William Roberts (1830-1899), a Manchester physician, performed extensive studies on organic solutions. Roberts received his training in physiology and medicine at University College, London. He served as Physician to the Manchester Royal Infirmary from 1855-83. In addition to presenting lectures on Anatomy, Physiology, Pathology, and Medicine, Roberts was the first Joint-Professor of Medicine at the Victoria University of Manchester from 1873-76. It was during this period that Roberts investigated alleged instances of spontaneous generation in the microbial world. In 1877 he became a Fellow of the Royal Society of London. Roberts was a most important influence upon both Ferdinand Cohn and John Tyndall during the events which comprise the Hay-Infusion Episode of the 1870's.

Writing in *Nature* in February, 1873, Roberts stated his belief that Bastian's results could be explained on the basis of technical errors. He suggested that Bastian's hermetic sealing had not been accomplished before a reflux of ordinary air had carried contaminants into his just-sterilized turnip-cheese infusions. Then too Roberts believed that Bastian's technique of heating infusion flasks directly over a flame caused the contents of the flasks to sputter during boiling. Thus bacteria-bearing droplets of the infusion adhered to the dry walls of the flasks and thereby escaped the destructive action of the boiling infusion. Upon cooling, when the flasks were moved about, these bacteria-containing droplets provided sources of contamination for reinoculation of the sterile infusion.[20]

The Turnip-Cheese Episode

Roberts also advocated the use of a water bath maintained at a temperature of 100°C. in which to submerge sealed vessels containing materials reputed to resist the ordinary sterilization procedure of simply boiling the infusion over a Bunsen burner. Using this technique, Roberts later found that at least twenty to forty minutes of the water-bath treatment were required to sterilize a turnip-cheese infusion.[21]

In June, 1873, Burdon-Sanderson published a detailed account of his own experiments on turnip-cheese infusion. In stating the purpose of his investigations it is clear that Dr. Roberts' initial remarks mentioned above had been noted by Burdon-Sanderson. The latter wrote:

> Certain particulars in Dr. Bastian's methods have been objected to as possible sources of uncertainty. Thus it has been suggested that when a flask, of which the neck has been drawn out to a capillary orifice, is boiled even for ten minutes over a lamp, it is not certain that the whole of the liquid contained in it is heated to the temperature of boiling; and again, that when the lamp is withdrawn in the act of closing the capillary orifice, germs may enter from without. Although I do not attach much importance to either of these objections, I have modified Dr. Bastian's method, so as to render them inapplicable. The modification, however, applies exclusively to the mode of heating the hermetically sealed flasks. As regards the preparation of the liquid, I have in no respects departed from his instructions.[22]

Burdon-Sanderson prepared his turnip-cheese infusion so that its specific gravity was 1018 to 1020. He filtered the infusion and neutralized it by sodic carbonate (sodium carbonate). In preparing the Cheddar cheese Burdon-Sanderson ground some cheese with a small amount of the turnip infusion, filtered this mixture through calico, and then added the filtrate to the containers of the infusion. When examined under the microscope every drop of the infusion contained bits of cheese.[23]

Burdon-Sanderson boiled and sealed some of his vessels of turnip-cheese infusion to confirm his observations of Bastian's work. With other boiled and sealed vessels, he differed from Bastian's procedures by using a Papin digester or pressure-cooker-like device, with which to apply an additional heat treatment. Burdon-Sanderson found that by boiling his flasks in a digester under the pressure of one inch of mercury he could apply a temperature of 100.92°C., under two inches of mercury a temperature of 101.81°C., and under three inches of mercury a temperature of 102.68°C.[24]

In his discussion of the temperature treatments to which he subjected his infusions, Burdon-Sanderson defined his terminology and made an astute observation with regard to bacteria.

> In describing the experiments I use the expression "turnip-cheese" liquid to denote the neutral infusion of turnip with cheese of which the preparation has been given above, and in recording the results the words barren and pregnant are employed to express the absence or presence of living Bacteria. In any liquid which has been kept five days at the temperature of fermentation there is no difficulty in determining in which of these two conditions it is, for if Bacteria are present at all they are present in such numbers that every field is crowded with them. Bodies which appear to be dead Bacteria are met with here and there in every specimen. They are as numerous in liquids examined immediately after prolonged boiling as in others. They are probably derived from the cheese.[25]

In his work Burdon-Sanderson was careful to inoculate a considerable number of his treated, barren flasks with non-sterile distilled water in order to "show that the liquid, although deprived of its power of germination, is as capable as before of supporting the life of Bacteria."[26]

After carefully tabulating his experimental results obtained by boiling turnip-cheese infusions under pressures greater than normal atmospheric pressure and hence at temperatures above 100°C., Burdon-Sanderson wrote:

The Turnip-Cheese Episode

> It is unnecessary for me to draw any inferences from the preceding experiments; it may not, however, be superfluous to point out that, although all the flasks heated above 101°C. remained sterile, this fact affords no ground for concluding that any definite relation exists between that precise temperature and the destruction of the germinating power of the liquid in question. All that has been shown is that the chance that such a liquid will breed Bacteria is diminished either by slightly increasing the temperature to which it is heated, or increasing the duration of the heating. Thus it appears to me quite probable that if a sufficiently large number of flasks were heated even to 102°C. some of them would still be found to be pregnant.[27]

Burdon-Sanderson's reluctance to establish a definite thermal death point for bacterial life indicates his insight into the very tentative state of microbial knowledge in England during the early 1870's. Furthermore, as late as Autumn, 1873, Burdon-Sanderson clung to his neutral position on the spontaneous-generation controversy. In October of that year he asserted that he did not consider "heterogenists as scientific heretics," but that he was "not aware of any proof that they are right."[28]

Edwin Ray Lankester and Dr. C.C. Pode carried out extensive investigations on turnip-cheese infusion throughout 1873. They believed that Bastian's results could be explained in several possible ways. In the first place, through hasty inspection microorganisms actually killed during boiling might be mistaken for living microbes. Furthermore, as Roberts had earlier suggested, sputtering might enable some bacteria to escape the full destructive action of the boiling infusion; or perhaps uneven heating of the turnip-cheese infusion vessels had been the cause of the presence of microbial life after boiling. Also, Lankester suspected that the cheese somehow protected bacteria from the full destructive effect of 100°C.

To obviate these difficulties, Lankester instituted the practice of microscopically inspecting the turnip-cheese infusion immediately after boiling as well as a few days after sealing. Smaller vessels, nearly filled to capacity with the infusion, were used. Lankester also employed a water bath similar to that used by Roberts. After five minutes boiling and sealing, he submerged the infusion vessels in boiling water for a period of fifteen to thirty minutes.

Finally, Lankester's concern for the role which the bits of cheese might play in Bastian's results was most interesting. He had referred to Pasteur's observations on the Butyric Bacterium and the resistance of that organism or of its germs to high temperatures. In his analysis of the importance of cheese in Bastian's results, Lankester mentioned Ferdinand Cohn's discovery that solid particles present in pea infusions rendered such infusions more heat resistant than filtered pea infusions.

In spite of his familiarity with the biological literature of his day, Lankester penetrated the turnip-cheese problem no more deeply than by suggesting mechanical explanations to account for Bastian's results. He seemed to feel that bacteria were physically protected from the destructive effects of both hot water and high temperatures by the poor absorptive qualities of cheese and by the insulating properties of that substance.

Lankester decided to carry out experiments on turnip infusions containing visible lumps of cheese in some instances and in other cases cheese in the form of an emulsion. Flasks containing visible particles of cheese frequently were not sterilized by being boiled and hermetically sealed and then submerged for five minutes in a bath of boiling water. However, flasks of turnip-cheese emulsion did remain free of microbial life after such treatment.

After performing many experiments on turnip-cheese infusion, Lankester concluded:

> No organic nor inorganic infusion has been contrived by Dr. Bastian nor by anyone else which will develop Bacteria, still less Torulae [yeast-like fungi], after exposure for one hour (or even less) to $212°F$. [$100°C$.]

The Turnip-Cheese Episode 35

Lankester left no doubt regarding his interpretation of turnip-cheese investigations when he wrote in October, 1873:

> It is my opinion that the only *positive* addition to knowledge which this inquiry about the development of Bacteria in infusions has led to is, that when you have cheese-emulsion, or similar material present in an infusion, you must be a little more careful about heating it than when you have not, if you wish to destroy by the agency of heat the life of Bacteria or their germs contained in the infusion. How it is that cheese emulsion helps the Bacterian contamination to escape destruction we do not know. Possibly in the same way as the larger lumps do. But that matter remains for inquiry when more is ascertained as to the natural history of Bacteria.[29]

Lankester's reference to the need for life-history studies of the bacteria was stated more clearly in his April, 1873, report on investigations carried out by him and by Dr. C.C. Pode at Oxford. In that report Lankester had written:

> Turnips and cheese may be very bad material for experiment; but it would be well, as far as possible, to settle the matter, or the way in which the matter is to be viewed with regard to them, before going off to other particular cases.
> It would be a very excellent thing if all further reference to this subject could be postponed for a year or two--that is, until further study of Bacteria, such as that inaugurated by Sanderson and Cohn, has given us surer ground to tread upon.[30]

Some answers to the turnip-cheese puzzle were to be found eventually through studies conducted by Ferdinand Cohn of Breslau in Germany.

Attempting to determine the thermal death point of bacteria, Lankester wrote that "it is desirable to recognize in our experiments the two distinct factors of this through-heating to any given temperature--namely, (1) the temperature to which the infusion to be heated is to be exposed; and (2) the length of time during which it is exposed to that temperature." He constructed a graph where "the perpendicular represents the range of temperature divided into degrees from 65°C. to 120°C." and where the horizontal axis is "divided into equal spaces representing periods of five minutes." "If the results of observations with a given infusion indicating the time of exposure to a particular degree of temperature required in order to prevent the subsequent development of Bacteria be marked off on such a scheme," he noted, "we should expect to obtain a series of points defining an asymptotic curve, the time required at the highest temperature being infinitely small, and at the lowest temperature infinitely great." He concluded that "this curve would vary in its character according to the properties of the infusion made use of." Lankester had elaborated upon Burdon-Sanderson's contribution of 1873 which pointed out that prolonged boiling at 100°C. could be replaced by brief exposures to temperatures in excess of the temperature of boiling water in order to achieve sterilization of materials reputed to resist high-temperature treatments. For example Lankester found that in the case of Bastian's turnip-cheese infusion an "exposure for *six hours* to *a temperature of 75°C. was sufficient* to prevent the subsequent development of Bacteria."[31]

In January, 1876, John Tyndall, Professor of Natural Philosophy at the Royal Institution in London, presented the results of his studies on organic solutions to the Royal Society of London. As a young man, John Tyndall (1820-1893) had journeyed from England to Germany with his friend, the chemist Edward Frankland (1825-1899) in order to study under Robert Wilhelm Bunsen (1811-1899) at the University of Marburg. After two years of intense effort Tyndall obtained his Ph.D. in 1850. In 1853 he became Professor of Natural Philosophy at the Royal Institution in London and in 1867 he succeeded Faraday as superintendent, a post which Tyndall held for twenty years. Involved in the spontaneous-generation controversy at the beginning of the 1870's, Tyndall continued to battle Henry Charlton Bastian throughout the decade of the 1870's.[32]

Tyndall had carried out extensive studies on pure liquid infusions, and he had found to his satisfaction that what he referred to as "five minutes boiling" sufficed to render these substances sterile. So adamant was Tyndall that he challenged Burdon-Sanderson: "I am therefore compelled to conclude that Dr. Sanderson has lent the authority of his name to results whose antecedents he had not sufficiently examined, and that the life to which he testifies, in the case of the purely liquid infusions, arose from errors of manipulation."[33]

On 10 February 1876 in *Nature* there appeared an article signed "Inquirer." The anonymous writer — very likely Henry Charlton Bastian — pointed to the conflicting findings of John Tyndall and John Burdon-Sanderson.

> I observe that Prof. Tyndall suggests that such contradictory results may be explained by "errors of preparation or observation." No doubt they may, but it would be a great shock to my scientific faith to be driven to this theory to explain apparent discrepancies between such observers as Professors Tyndall and Sanderson. I cannot help, not only hoping, but believing that there must be some way of reconciling the experiments of two such eminent inquirers, and I should be much perplexed if I were compelled to form an opinion whether the supposed error, if it does exist, ought to be attributed to the one or the other.

"Inquirer" also observed that, although Tyndall had performed a great many more experiments than Burdon-Sanderson's single investigation on turnip-cheese infusion, his exact procedures and techniques had not been described in as great detail as Burdon-Sanderson's work.[34]

The explanation of the difference between Tyndall's results and those obtained by Burdon-Sanderson was to be given by Ferdinand Cohn. Trained as a professional botanist at Breslau and at Berlin under Johannes Müller (1801-1858), Ferdinand (Julius) Cohn (1828-1898) accepted a position at Breslau in 1859. Since the early 1850's he had carried out detailed studies of the life histories of algae and fungi. As

Director of the Plant Physiology Institute at Breslau, Cohn concentrated upon the bacteria during the 1870's. In addition to establishing a system of classification for the bacteria, Cohn placed much emphasis upon working out the life cycles of individual bacteria. His work led to the discovery of endospore formation in *Clostridium butyricum* during the summer of 1873 and in *Bacillus subtilis* by July, 1876. Cohn was a very kind man who enabled the young country doctor Robert Koch to publish his paper on the causal organism of anthrax, *Bacillus anthracis*, in spring, 1876. It is unfortunate that Cohn has not received his proper share of credit for the breakthroughs in microbiology which characterized the decade of the 1870's.[35]

Men such as Burdon-Sanderson, Lankester, Roberts, and Tyndall tended to focus upon the characteristics of the organic infusions with which they were dealing. In the early 1870's none of these men knew much about the biology of microorganisms and indeed the behavior of bacteria was equally mysterious. Burdon-Sanderson described the microorganisms which he observed in Bastian's boiled and sealed turnip-cheese infusion as motile bacteria of moderate size which formed long filaments of bacterial cells attached end to end. He also mentioned granules which strongly refracted light. Furthermore, Lankester had referred to Pasteur's observation of the heat resistance demonstrated by the butyric-fermentation organism or by the germs of that bacterium. Nevertheless, British investigators concentrated primarily upon mechanical explanations of heat resistance. But Cohn's background of investigating the life histories of algae and fungi and his practiced hand and eye applied to microscopes of at least 2,000-diameter magnification, allowed him to approach the turnip-cheese episode quite differently from his counterparts in England.

The German botanist had been wrestling with the problem of microbial durability prior to Burdon-Sanderson's certification of Bastian's results and even before he had received news of Bastian's turnip-cheese infusion.[36] Lankester had referred to Cohn's discovery that pure liquid pea infusions were more readily sterilized by boiling and sealing than were pea infusions containing even a single pea in the otherwise pure liquid infusion. But Cohn's remarks on the difficulty of rendering certain kinds of organic materials sterile by subjecting them to the temperature of boiling water did not stop here, for he observed that a frequent type of living microorganism in such boiled, but non-sterile, organic solutions was a bacterium which he classified with the rod-shaped bacteria in the genus *Bacillus*.[37] He also knew about Pasteur's 1861 description of the microorganism responsible for butyric fermentation,[38] and he was struck by the similarity between Pasteur's butyric microbe and his own bacillus organism. Well acquainted with Pasteur's 1862 detailed account of the failure of temperatures up to 105° C. to sterilize milk, Cohn was aware of Pasteur's conviction that this unusual degree of heat resistance could be attributed to the germs of a microorganism. He was again impressed by the parallel between his own *Bacillus* bacterium and the microbes discussed by Louis Pasteur in connection with the heat resistance of milk.

Although he was not dogmatic on this point, Cohn was fairly convinced of the likelihood of spore formation in the bacteria by 1872. From his extensive studies of the life histories of the algae and fungi, Cohn had good reason to be watchful for the occurrence of analogous resting bodies in the bacteria.[39] Thus in May, 1873, Cohn turned his attention not to the characteristics of the infusion involved, but rather to the kinds of microorganisms which might appear in such an infusion.[40]

During the summer of 1873 Cohn had an opportunity to study the manufacture of cheese in great detail. The process of cheese production results from the biochemical activity of many different kinds of microorganisms.[41] His investigations of rennet infusion which forms the base for cheese production revealed to Cohn some rod-like bacteria which closely resembled Pasteur's butyric microbe and indeed appeared similar morphologically to his own *Bacillus subtilis*. Furthermore, Cohn noticed the globular swellings at the terminal ends of individual bacillus cells which we know today to be the characteristic manner of spore formation in the genus *Clostridium* which contains anaerobic bacteria (bacteria living in the absence of free oxygen) which carry on butyric-acid fermentation.

Cohn therefore explained that spore-forming bacteria of the bacillus type, present in rennet infusion essential for cheese production, became encased in the cheese as it dried and hardened. Whereas the adult bacilli eventually perished within the dry cheese, the durable resting bodies or spherical spores formed at the terminal ends of the individual bacillus cells survived within the cheese. With the return of favorable conditions, these spores germinated and grew into new adult bacilli or rod-like bacteria.

In Bastian's neutral turnip infusion with bits of Cheddar cheese dispersed throughout it, the heat-resistant spores of rod-like bacteria survived the five to ten minutes boiling and sealing treatment. After the flasks had cooled sufficiently, these bacillus spores, formerly encased in the bits of cheese, germinated in the nearly anaerobic conditions of the hermetically sealed flasks and developed into adult bacilli nourished by the very suitable turnip infusion.[42]

A succinct statement of the results obtained by Cohn in his investigations on turnip-cheese infusion appeared in the 29 June 1876 issue of *Nature*.

> The formation of cheese has lately engaged the attention of Prof. Ferd. Cohn in connection with his researches on the lowest forms of plant life; and he has made personal observations on the manufacture, as carried on in Switzerland. The phenomena accompanying the process are thus described: The rennet contains a liquid ferment which causes coagulation of the milk; also ferment-organisms (Bacillus) which probably bring on

butyric-acid fermentation, and cause the slow maturing of the cheese. It is their resting-spores that, enclosed by the dry cheese substance, resist boiling heat for a long time, and, in a suitable nutritive liquid, may afterwards develop to bacillus rods. (One of Dr. Bastian's results is thus explained.)[43]

In Volume II of his 1897 *Die Pflanze*, Ferdinand Cohn devoted one paragraph and one illustration to the spore-forming bacterium responsible for Bastian's puzzling results with neutralized turnip-cheese infusion. In that botanical work of 1897, Cohn assigned the troublesome microbe the designation *Bacillus (Clostridium) butyraceus*,[44] although today the microbe of the Turnip-Cheese Episode is generally termed *Clostridium butyricum*.

Turnip-cheese infusion investigations were an important step in man's effort to clarify the nature of bacterial germinal matter and to establish a thermal death point for bacterial spores in the decade of the 1870's. The empirical attempts of British investigators to answer the enthusiastic claims of the heterogenist Henry Charlton Bastian had attracted the attention of the man best qualified in the early 1870's to solve the mysteries of heat resistance demonstrated by various organic solutions. That man, Ferdinand Cohn, became the Founder of Bacteriology. Through emphasis upon life-history studies of individual microorganisms, Cohn was able to account for alleged occurrences of microbial spontaneous generation in the 1870's.

If British workers helped to draw Cohn into the dispute over the *de novo* origin of microbes and into some of the most brilliant detective work in the history of microbiology, Cohn was not long in making an impact of his own. Three months after the notice in *Nature* concerning his experiments on cheese manufacture and his demonstration of the presence of heat-resistant bacillus spores in Bastian's turnip-cheese infusion, Cohn paid a visit to London in September, 1876, and handed to John Tyndall copies of his own paper on *Bacillus subtilis* and that of Robert Koch on *Bacillus anthracis.* These papers of both Cohn and Koch were to have far-reaching implications upon man's study of the biology of microorganisms in England.

Chapter 4
The Hay-Infusion Episode

In January 1873 John Scott Burdon-Sanderson published a report confirming results obtained by Henry Charlton Bastian. Bastian had pointed out the frequent failure of five or ten minutes' boiling at 100°C. followed by hermetic sealing toward the end of ebullition to sterilize organic solutions.[1] Thus Burdon-Sanderson signaled the beginning of two major episodes in the disputes over spontaneous generation during the decade of the 1870's: turnip-cheese studies, already examined, and hay-infusion investigations.

Burdon-Sanderson commented upon the microbes which subsequently appeared in hay-infusion flasks which had been boiled for five minutes and then hermetically sealed toward the end of ebullition. The hay infusion employed in Bastian's experiments, which had been observed by Burdon-Sanderson, possessed a specific gravity of 1005 and was of neutral reaction. Six days after the organic solution had been prepared and treated by five minutes' boiling and hermetic sealing, Burdon-Sanderson described one of Bastian's hay-infusion vessels. Over a three-day period, the flask had become quite turbid. Yet the hermetic seal on the flask was intact. Once it was opened and the liquid inside examined, Burdon-Sanderson found actively-moving, minute bacteria. There were also many colonies of spherical particles giving rise to bacteria. Furthermore, threads of bacterial cells were present in the fluid.[2] Burdon-Sanderson's description of the microbes which subsequently appeared in Bastian's treated hay-infusion flasks bears a remarkable resemblance to the masterful essay on this subject to be written more than three years later by Ferdinand Cohn.[3]

In England the Manchester physician William Roberts carried out investigations on the effect of elevated temperatures upon a great variety of organic solutions and especially upon hay infusions, and he had attempted in 1873 to account for Bastian's results by suggesting possible sources of error in the London professor's procedures. In Roberts' report to the Royal Society of London on 16 April 1874, however, he also had clearly encountered amazing instances of heat resistance among various types of infusions. Especially troublesome were organic solutions prepared from hay.[4]

Roberts had found that an organic solution made from freshly harvested hay was readily sterilized by five minutes' boiling. If, however, the hay infusion had been made slightly alkaline prior to boiling and sealing, then in some cases it could not be sterilized by less than three hours of boiling at $100°C$. Roberts had studied Bastian's work on organic solutions, Burdon-Sanderson's investigations on microbes, and Ferdinand Cohn's 1872 report on bacteria.[5]

Roberts' conclusions in his report of 1874 testify to his profound insight into the puzzling problems of heat resistance in the microbial world. For example, he commented upon the remarkable difference in heat resistance between acidic as compared to alkaline hay infusions, and he referred to possible differences in heat tolerance between adult bacteria and the "germs" from which they possibly arose. Roberts again emphasized: "We have seen that unneutralized hay-infusion is sterilized by exposure to a boiling heat for five minutes; but if the infusion be slightly alkalized by ammonia or potash, it germinates after exposure to the same heat for more than an hour."[6]

Roberts' excellent paper received careful consideration in England and also in Germany. By September 1875, John Tyndall was engaged in a detailed examination of the effect

The Hay-Infusion Episode 45

of five minutes' boiling upon purely liquid organic solutions. In Germany, Ferdinand Cohn began his study of hay infusions on 25 October 1875. Tyndall's results were initially published in January 1876;[7] Cohn's results appeared in July 1876.[8]

Tyndall's summary of his research conducted during autumn and winter, 1875, stated that in all instances of purely liquid organic infusions examined by him five minutes' boiling followed by hermetic sealing toward the end of ebullition sufficed to render these solutions free of microbes whether they were of neutral, acidic, or alkaline reaction.[9] A significant feature of Tyndall's report of January, 1876, was that British physicist's serious questioning of the accuracy of Burdon-Sanderson's and Roberts' work since their findings conflicted with Tyndall's results.

Commenting on Burdon-Sanderson's January 1873 paper, Tyndall noted that his own infusions were of the same specific gravity as those used by Bastian. He also took special care in studying the infusions used in Bastian's investigations which Burdon-Sanderson had validated. Tyndall reiterated that Bastian had convinced Burdon-Sanderson that subsequent to boiling and hermetic sealing bacteria did appear in at least some of Bastian's infusion flasks. Tyndall insisted: "I am therefore compelled to infer that the instances in which Dr. Bastian failed to obtain Bacteria in his hermetically sealed tubes were illustrations of correct experimenting, the appearance of life in other cases being the result of errors of manipulation."[10]

Tyndall's preoccupation with experimental errors as an explanation for the discrepancy between his own results and those obtained by other investigators is further illustrated by Tyndall's January 1876 discussion of Roberts' work. William Roberts strongly supported the germ theory; however, he stressed the necessity of leaving open the possibility of microbial spontaneous generation in rare instances. Tyndall focused on Roberts' use of a dry cotton-wool plug interposed between the bulb of a pipette and the extremity of its neck. After hermetic sealing, the bulb was then subjected to boiling in a water or oil bath for the desired time. Following ebullition, Roberts removed the bulb and allowed the vessel to cool. He then filed off the neck of the bulb, leaving only the cotton wool as a barrier or filter between the infusion and the atmosphere. Although Tyndall complimented Roberts on his ingenious arrangement, he insisted that it was not possible to find cotton wool free of germs. He further contended that temperature variations, air movement, and even slight jarring of the pipette would be sufficient to dislodge germs from Roberts' plugs. Tyndall also asserted that condensation in the pipette necks might also provide an avenue for germs to be carried from the plug downward into the infusion contained in the bulb. Since only occasional bulbs showed evidence of microbes after boiling, Tyndall believed that Roberts' careful techniques were responsible for the many instances where vessels were successfully treated and remained sterile. Tyndall noted that Roberts had left the possibility of microbial spontaneous generation open since some of his infusion vessels contained living microbes after long periods of heat treatment. He also mentioned Roberts' finding that alkalized infusions of hay could not be sterilized even after several hours of boiling. Referring to his own work on this infusion, Tyndall insisted that "five minutes' boiling sufficed to completely sterilize the liquid." But Tyndall was willing to grant that "Dr. Roberts is certainly correct in assigning to it superior nutritive power." His appraisal of Roberts' views regarding spontaneous generation appears to be a little severe, for they were very tentative and carefully expressed.[11]

Tyndall's report of early 1876 and especially his remarks concerning Burdon-Sanderson and Roberts received quick notice in England and also in Germany. In England Bastian in February 1876 pointed out the discrepancy between Burdon-Sanderson's report of 1873 and that of Tyndall in 1876.[12]

The Hay-Infusion Episode

Tyndall's letter of reply, published in *Nature* in February (1876), suggested that Burdon-Sanderson repeat his experiments on purely liquid infusions, as opposed to mixtures of solids and liquids. He also suggested that if Burdon-Sanderson were to do so he would probably be led to the "right result" which, it is clear, would be in agreement with Tyndall's findings.[13]

That Burdon-Sanderson very much resented Tyndall's hasty judgment of his infusion work is readily apparent from his paper entitled "Remarks on the Attributes of the Germinal Particles of *Bacteria*, in Reply to Prof. Tyndall,"[14] which was published in *Nature* nearly two years after Tyndall's initial comments on Burdon-Sanderson's work. More will be said of this later.

Concerning William Roberts' efforts, there is evidence that Tyndall reevaluated them later. Writing in the spring of 1876, Tyndall somewhat modified his opinion of Roberts' work. Having restricted himself to the study of purely liquid infusions, he indicated that his next investigations would focus on milk, turnip juice and cheese, peptone, and "the remarkable experiments described by Dr. William Roberts, a small residue of which only I have failed to corroborate." [15] This was quite an admission on Tyndall's part when one recalls his earlier, adamant stand on "five minutes' boiling" as an effective agent of sterilization for all kinds of infusions.

Throughout the winter and early spring of 1876, however, Tyndall's view of Burdon-Sanderson's work remained unchanged and in fact in his April 6th report Tyndall expressed himself even more strongly regarding the possibility that Burdon-Sanderson might have been deceived in his support of the accuracy of Bastian's experiments on hay infusion and on turnip-cheese infusion. Tyndall again noted that Bastian had convinced Burdon-Sanderson that bacteria could occasionally appear in vessels after boiling and hermetic sealing. He reiterated his own inability to duplicate such experiments with purely liquid infusions made from hay and from turnips. In repeated trials of heating and sealing, Tyndall found that such liquid infusions always remained sterile. He made the rather severe accusation that Burdon-Sanderson had not scrutinized Bastian's work carefully enough and that the British physiologist had been misled by "errors of manipulation."[16]

In Germany meanwhile Ferdinand Cohn was also studying hay infusions. Like Tyndall, he had read Bastian's work and Roberts' 1874 paper. On 25 October 1875, Cohn began his investigations on organic solutions prepared from hay.[17] He had already undertaken an examination of possible explanations for the heat resistance of turnip-cheese infusion on 9 May 1873.[18] That series of researches had led him during the summer of 1873 to the discovery of spore formation in the rod-like bacterium, now designated *Clostridium butyricum*, a microbe important in the production of Cheddar cheese.[19]

With more than twenty years of experience in studying the life histories of minute algae and fungi, Cohn's approach to the alleged heat resistance of alkalized hay infusions would differ markedly from the methodology employed by the British physicist John Tyndall. He set out to determine the kind of living microorganism which appeared in hay-infusion vessels which had been boiled for five minutes and then hermetically sealed toward the end of ebullition. Invariably, microbes found alive in vessels a few days after treatment by Tyndall's supposedly lethal methods were rod-like bacteria termed by Cohn *Bacillus subtilis* or hay bacilli.[20] Then Cohn began a detailed study of that organism. He observed the formation of endospores within individual bacillus cells — these cells frequently being united end-to-end and thereby forming long filaments or Leptothrix threads. (Burdon-Sanderson had mentioned such threads in early 1873.)[21]

Although Cohn corroborated William Roberts' findings regarding instances of amazing heat resistance among flasks of hay infusions, he failed to observe the distinct correlation between alkaline hay infusion and superior heat resistance which Roberts had recorded. Rather, Cohn believed that the characteristics of the hay itself determined the kind of heat treatment required to kill the microbes in an organic solution made from hay.[22]

In his July 1876 report, Cohn attempted to explain the discrepancy between Tyndall's results of January 1876, and those of Roberts in April 1874. He suggested that Tyndall's use of an oil bath during the boiling and hermetic-sealing process had raised his hay-infusion flasks to a temperature above 100°C. and thus rendered them sterile, whereas Roberts' vessels placed in water baths had been brought only to the temperature of boiling water.[23] Cohn had made an incorrect inference in this regard. As Tyndall was to point out nearly a year later, since his own flasks were unsealed, they had not been heated appreciably above 100°C.[24] Cohn's misconception concerning Tyndall's treatment of organic solutions, however, probably arose from the fact that by 1876 he well knew the ineffectiveness of 100°C. to sterilize pea, turnip-cheese, and hay infusions. Being misled by Tyndall's use of an oil bath during the sterilization process, Cohn apparently had jumped to a possible explanation to reconcile Tyndall's results with those obtained by William Roberts.[25]

From April 30 — May 3, 1876, Robert Koch visited Ferdinand Cohn in Breslau to demonstrate the entire life history of the bacterium which is the causal organism of anthrax — *Bacillus anthracis*.[26] Well trained in medicine and also in the sciences under Friedrich Gustav Jakob Henle (1809-1885) at the University of Göttingen, Robert Koch (1843-1910) earned his medical degree in 1866. In 1870-71 he was an army surgeon in the Franco-Prussian War. After the war he began his medical practice near Breslau, and it was during this period that he began his work on anthrax which led to revolutionary changes in medical microbiology. Koch and his students developed many of the modern techniques used daily in the microbiological laboratory such as the use of solid nutrient media to culture bacteria and efficient methods of preparing bacteria for study by means of fixing and staining procedures.[27] Cohn's work on the hay bacillus, *Bacillus subtilis*, and Koch's investigations of the anthrax pathogen paralleled one another. Both microbes form endospores which are highly resistant to adverse environmental conditions such as extreme dryness and high temperatures. Cohn was instrumental in getting Koch's work published, and his July, 1876, paper on *Bacillus subtilis* devoted its final section to introducing Robert Koch's essay on the anthrax bacillus.[28]

The papers of Cohn and Koch were soon to make a distinct impression on John Tyndall in England. Corroboration of William Roberts' experiments on the resistance of hay infusion to five minutes' boiling and hermetic sealing by Cohn definitely played a part in stimulating Tyndall to renew his study of organic solutions in autumn, 1876. In late September a social meeting occurred in London between Ferdinand Cohn and John Tyndall;[29] perhaps this was the occasion on which Cohn gave a copy of his journal to Tyndall. The second phase of Tyndall's infusion experiments reflects Cohn's influence and indicates that Tyndall had read Cohn's paper on *Bacillus subtilis* by mid-October. Tyndall himself later wrote: "On my return from Switzerland last autumn [fall of 1876] the experiments on alkalized hay-infusions were resumed; and soon afterwards Professor Cohn, of Breslau, so highly distinguished by his researches on *Bacteria*, placed in my hands a memoir which rendered it doubly incumbent on me to examine more strictly the grounds of my dissidence from Dr. Roberts."[30] (Tyndall also referred to Koch's work on anthrax in his Glasgow Address on 19 October 1876.[31])

In response to Cohn's paper on *Bacillus subtilis*, Tyndall brought dried hay into his Royal Institution laboratory and began to prepare infusions from that desiccated material. By November and December, Tyndall found that not only his

The Hay-Infusion Episode 51

hay infusions but other purely liquid vegetable and animal infusions could not be sterilized by five minutes' boiling as they had been in autumn and winter, 1875. In fact, microbes subsequently developed in some of Tyndall's hay infusion vessels which had been boiled for several hours and hermetically sealed toward the end of ebullition. In one instance, Tyndall observed that even eight hours of boiling did not sterilize hay infusion.

Summarizing his investigations of autumn, 1876, Tyndall observed: "In relation to our present experiments, the influence of drying and hardening was brought home to me by the fact that in all the foregoing cases the infusions which five minutes' boiling proved sufficient to sterilize *were, without exception, derived from fresh hay mown in 1876, while the infusions which five minutes' boiling failed to sterilize were derived, without exception, from old hay mown either in 1875 or some previous year.*[32] In the first few days of January, 1877, Tyndall moved from the Royal Institution laboratory in order to seek a less contaminated environment at the newly-opened Jodrell Laboratory at Kew Gardens. Here Tyndall planned to resume his investigations on organic solutions in an effort to obtain once again results similar to those recorded in autumn and winter, 1875.[33]

At Kew Gardens Tyndall was able to duplicate his work of the previous year. He therefore returned to the Royal Institution in an attempt to repeat his Kew experiments in a special workroom which he had constructed on the roof of the Royal Institution. After many hardships, he was finally able to perform experiments on a variety of infusions which provided results very similar to those obtained in 1875 and at Kew Gardens in early 1877.[34]

On 14 February 1877 in a letter written to Thomas Henry Huxley (then Secretary of the Royal Society of London), Tyndall described his method of discontinuous heating, a technique he had developed during his work on hay infusions.[35] Discontinuous heating or Tyndallization will be discussed later.

Chapter 4

In May 1877 Tyndall published a complete report of his experiments with hay infusions throughout the previous autumn and winter. Correcting Ferdinand Cohn's misconception published in July of the previous year, relating to Tyndall's use of an oil bath which Cohn had inferred caused Tyndall's infusion vessels to be raised above 100°C., Tyndall explained that his vessels had not been sealed and hence were heated only to the boiling points of the infusions.[36] Tyndall's paper also revealed its author's new respect for the researches of William Roberts on alkaline hay infusion and their bearing on microbial thermal resistance.[37] But no reference can be found in Tyndall's paper to John Burdon-Sanderson's report of January, 1873.

If Tyndall had permitted himself to forget his remarks concerning Burdon-Sanderson's work on organic infusions, the British physiologist did not share the same shortness of memory. In *Nature* for 29 November 1877 Burdon-Sanderson published a paper entitled "Remarks on the Attributes of the Germinal Particles of *Bacteria*, in Reply to Prof. Tyndall." After recalling Tyndall's unfavorable comments made in the *Philosophical Transactions* for April, 1876, Burdon-Sanderson noted that Tyndall had suggested that he repeat Bastian's experiments which he had certified. Having performed his own experiments on turnip-cheese infusion, Burdon-Sanderson restated his conclusion that "the proneness of the liquid to produce *Bacteria* can be diminished either by increasing the temperature employed to sterilise it, or if the ordinary temperature of ebullition be used, by prolonging its duration." He also explained that he had not pursued his infusion work after 1873 because he had demonstrated that "as a ground for believing in spontaneous generation the turnip-cheese experiment was a failure" and that "the subject had been taken up by the most competent living observers, who had in every particular confirmed the accuracy of my results." Finally, Burdon-Sanderson wrote that his only reason for speaking out was to show that "the

The Hay-Infusion Episode 53

statements which Dr. Tyndall in 1876 characterised as incautious, and which he virtually invited me to retract, had been two years before confirmed in every particular by experimenters of acknowledged competence."[38]

Thus John Tyndall was forced to revise his estimates of William Roberts' work not only by his own study of infusions in spring and summer, 1876, but especially by Ferdinand Cohn's corroboration of Roberts' findings which Tyndall received and definitely read in autumn of that year. Regarding Burdon-Sanderson's investigations on the heat resistance of hay and turnip infusions, Tyndall apparently did not modify his views of that British physiologist's researches on organic solutions, but in print Burdon-Sanderson came to his own defense in an article published nearly two years after Tyndall's initial comments.

Concerning John Tyndall's relationship with Ferdinand Cohn, Tyndall's subsequent references to him made little or nothing of the fact that his paper on *Bacillus subtilis* had played a very significant part in placing Tyndall on the road to grasping the full significance of bacterial endospore formation. From there Tyndall devised his method of discontinuous heating to cope with cases of unusual thermal resistance. But Tyndall's tributes to Robert Koch appear to be plentiful and very generous. Perhaps Tyndall, an advocate of the germ theory, considered Koch's work on anthrax to be more pertinent than Cohn's investigations on minute forms of plant life.

One illustration of a distinct absence of any expression of indebtedness to Cohn may be found in Tyndall's essay entitled "Spontaneous Generation" (1878). Only one reference is made to Cohn, and it in no way concerns the German botanist's discovery of bacterial endospore formation. Also, there are at least four places in Tyndall's essay where it would have been possible or even very logical for the British physicist to have paid some tribute to Cohn. As is frequently the case with Tyndall's writings after the discovery of spore formation in bacteria, Robert Koch and the anthrax pathogen are distinctly discussed.[39]

In a similar manner, an essay written by John Tyndall to serve as an introduction to the English translation of Louis Pasteur's biography refers to Koch at length, but it fails to acknowledge the help which Tyndall undoubtedly received from Cohn's paper on *Bacillus subtilis* which he had in his possession and had definitely read in autumn, 1876.[40] A final illustration of Tyndall's substantial discussion of Koch's contribution with only passing reference made to Cohn is found in another statement attributed to Tyndall in 1885.[41]

Studies on the resistance of hay infusion to conventional sterilization techniques lasted for over four years and absorbed the interest of Bastian, Burdon-Sanderson, Roberts, Tyndall, and Cohn. During much of this phase of the spontaneous-generation debate, Tyndall had insisted that "five minutes' boiling" was sufficient to sterilize organic solutions. Attempting to establish the germ theory of disease in England, Tyndall regarded the heterogenist Bastian as his principal opponent throughout the 1870's. Yet unwittingly, Tyndall became a supporter of Bastian's misplaced confidence in the lethal effect of $100°C$. upon all microbes. Disparaging the work of Roberts and Burdon-Sanderson for a time, Tyndall was stimulated to resume his hay-infusion investigations in autumn, 1876, after he had read Ferdinand Cohn's paper on *Bacillus subtilis*. His new series of experiments made Tyndall appreciate the extreme durability of bacterial endospores and enabled him to devise his sterilization method known as discontinuous heating or Tyndallization.

Chapter 5
The Urine Episode

In autumn 1875, the London heterogenist Henry Charlton Bastian began to conduct experiments on urine in an effort to support his belief that under certain conditions microbes could arise *de novo* from non-living, organic solutions. Bastian claimed that acidic urine boiled for five minutes at 100°C. could be made to swarm with bacteria provided that the urine was neutralized with boiled potash solution and then stored for a few days at 50°C. He regarded five minutes' boiling at 100°C. to be an effective sterilization procedure for acidic as well as neutral liquids. In other words, Bastian insisted that 100°C. was lethal for all microbial germs.[1] At this time John Tyndall, a germ theorist and an opponent of Bastian, likewise considered five minutes' boiling at 100°C. sufficient to sterilize all liquids. Tyndall did not begin to modify his stand regarding five minutes' boiling until autumn and winter, 1876.[2] Bastian's investigations eventually led to a vigorous debate over the response of urine to high-temperature treatment and to a meeting with Louis Pasteur in Paris in mid-July, 1877.

By July 1876, Bastian's claims that microbial spontaneous generation could be made to occur in boiled and subsequently neutralized urine had been noted by Pasteur. Educated in Arbois, in Besançon, and in Paris at the *École Normale Supérieure*, Louis Pasteur (1822-1895) had already earned a solid reputation as a French chemist when in 1857 he became director of scientific studies at the *École Normale Supérieure*.

It was at this time that Pasteur's attention turned to biology and led to his brilliant researches on fermentation and spontaneous generation. In recognition of his efforts to establish the germ theory of disease Pasteur was made a member of the French Academy of Medicine in 1873. In the latter half of the 1870's he worked on anthrax and extended the investigations of Robert Koch by developing an anthrax vaccine which was proved to be efficacious at a public demonstration in 1881. After intensive study of rabies and the development of a cure for that dread disease, Pasteur worked at the Pasteur Institute, opened in 1889, until his death just six years later.[3]

Pasteur readily agreed with Bastian that 100° C. applied to acidic urine was in his opinion a suitable treatment to prevent the appearance of microbes in the urine following boiling and sealing. Indeed it was simple enough to boil acidic urine vessels and to observe their barrenness for prolonged periods of time. The crucial issue between Bastian and Pasteur was the neutralization process which occurred after acidic urine had supposedly been sterilized. While Bastian contended that five minutes' boiling at 100° C. was sufficient to sterilize potash solution, Pasteur had long recognized that neutral and alkaline solutions resist the sterilization action of ordinary boiling. As early as 1862 the French investigator had published his findings that milk required 105° C. or 110° C. to be rendered free of microbial life.[4] Similarly, William Roberts in spring, 1874, had stressed the difficulty of sterilizing alkalized hay infusion as compared to acidic hay infusion. In Pasteur's view microbial germs managed to resist the destructive action of high temperature far more effectively in neutral or alkaline solutions than in acidic ones. Thus Pasteur regarded Bastian's neutralization process as the avenue by which microbial germs gained entry to the previously sterilized acidic urine, and after sufficient time for development, gave rise to a noticeable crop of bacteria in the now neutral urine.

The Urine Episode

If only the agent of neutralization were truly free of microbes before its addition to the supposedly sterile acidic urine, Pasteur believed that Bastian's arguments on behalf of the generating power of urine would crumble. To begin his attack against Bastian's claims, Pasteur boiled acidic urine for five minutes at 100° C. and then added a bit of solid potash to the urine, the fused potash having been previously heated to glowing redness. He then stored the urine flask at 50° C. and observed that it remained free of microbes indefinitely. Next, Pasteur repeated his experiment with one modification. To quiet Bastian's criticism of his procedure, this time he used a potash solution as the neutralization agent. Rather than merely boiling it for five minutes at 100° C. as Bastian had done, however, Pasteur subjected the potash solution to severe heat treatment in order to insure its sterility. In one instance, he heated the potash solution to 110° C. for twenty minutes; on another occasion he subjected the potash solution to 130° C. for five minutes. Neutralization of boiled acidic urine with potash solution treated by Pasteur's method yielded no microorganisms even after the urine had been stored at 50° C. for extended periods of time. These preliminary exchanges between Bastian and Pasteur during summer, 1876, merely had the effect of defining the issues of the urine battle.

The position taken by Pasteur in the urine episode had two strong English supporters. These opponents of heterogenesis were William Roberts and John Tyndall, whose experimental results on urine were presented to the Royal Society of London on 21 December 1876. During the previous summer, bacterial endospores had been studied and described in Germany by Ferdinand Cohn and Robert Koch. News of these highly-resistant germs had reached England in autumn when Cohn visited John Tyndall in London and gave him copies of the famous papers on *Bacillus subtilis* and *Bacillus anthracis.* Autumn 1876 also marked the beginning of an important modification of Tyndall's views. Experiments conducted by that British physicist led him in January, 1877, to renounce his long-held view that five minutes' boiling at 100°C. was an effective agent for sterilizing organic solutions.

William Roberts' paper[5] mentioned that Bastian had observed that "whether the alkali was added before or after ebullition he obtained the same result — the urine in both cases became fertile; and he concluded that the alkali had a positive power of promoting the origin of organisms in the urine." Roberts selected for consideration the two central points of the entire debate over microbial spontaneous generation in neutralized urine, namely that "it must be ascertained beyond doubt that the boiled acid fluid has been really deprived of its germs" and that "in adding the liquor potassae due care is taken that no new germs are introduced at the same time." In an attempt to fulfill the first of these conditions, Roberts designed his experiments so that "the acid infusion, after it had been boiled, was set aside in a warm place for a fortnight in order to test its sterility" and "the liquor potassae was not added to it until the lapse of time had satisfied me that it had been rendered permanently barren." To insure the sterility of the potash solution, he placed a sealed vessel of liquor potassae in an oil bath and heated it to 280°F. [137.8°C.] for fifteen minutes. Bastian had used just enough potash solution to neutralize an ounce of normal urine.

On 21 December 1876, John Tyndall also described his own experiments on the question of microbial spontaneous generation in the case of boiled and subsequently neutralized urine.[6] He had heated little tubes containing specific quantities of caustic potash to 220°F. [104.4°C.] for fifteen minutes. Inserting these tubes into urine flasks, he then boiled the urine vessels for five minutes and sealed them. He next set aside the urine flasks in a warm area to allow time for any contaminants to show evidence of microbial growth. Thereafter, each urine vessel was shaken so as to break the small potash tube which it contained. Incubating his now neutralized urine flasks at 122°F. [50°C.] to provide optimum conditions for microbial growth, Tyndall reported that after two months not one of his ten urine flasks had undergone any change resembling microbial spontaneous generation. He expressed doubt that such a phenomenon would ever take place in his ten flasks. Commenting upon the similarity of the results obtained by him and by William Roberts, Tyndall pointed out that while Roberts' potash vessels had been heated to 280°F. [137.8°C.] his own potash solution had been treated at the much lower temperature of 220°F. [104.4°C.]. Tyndall also reminded the Royal Society that during the previous summer Pasteur had conducted experiments similar to those of Roberts and Tyndall and had obtained results contrary to those which Bastian observed in his experiments. He remarked that it was Pasteur who, long before Roberts and Tyndall, had subjected potash to temperatures above 100°C. and had found that no microbes developed in neutralized urine. Pasteur had heated solid potash to redness. He had heated potash solution to 110°C. Neutralization of the urine by Pasteur's methods never led to the development of microbes.

A little more than a month after Roberts' and Tyndall's experiments on urine were presented to the Royal Society of London, Pasteur addressed the French Academy of Sciences. He offered a definite challenge to Bastian in an effort to bring an end to the urine controversy. On 29 January 1877, Pasteur defined Bastian's position and his own stand against that position. He stated that Bastian had found living bacteria in a vessel of neutralized urine held at 50°C. Bastian had added boiled potash solution to supposedly sterile urine and had obtained results which he regarded as revealing "the physico-chemical conditions of the spontaneous generation of certain bacteria." Pasteur challenged Bastian to repeat his alleged spontaneous-generation experiment, this time using pure potash and pure water. If Bastian wished to use impure potash solution, then such a solution could be made from the English or some other Pharmacopoeia provided that the potash solution, whether strong or weak, was subjected for twenty minutes to 110° C. or for five minutes to 130° C.[7]

On 12 February 1877, a note from Bastian was read to the French Academy of Sciences in which the London heterogenist accepted Pasteur's challenge. Bastian stated that he had scrutinized his experiments far beyond Pasteur's recommendations. He reported that his potash solution in one instance had been held for sixty minutes at 110° C. and in another instance for twenty hours at 110° C. As when the potash solution had been heated to only 100° C. and then added to supposedly sterile urine in appropriate amounts, Bastian still found that *"in twenty-four to forty-eight hours the urine was in full fermentation and swarmed with bacteria."* He wrote that his various urine samples possessed a specific gravity of 1020 to 1022 and were neutralized by 3% liquor potassae. Urging Pasteur to admit defeat in the controversy over microbial spontaneous generation in neutralized urine, Bastian expressed his willingness to perform his investigations "before competent judges."[8]

Pasteur immediately replied to Bastian in that meeting of 12 February 1877. He thanked Bastian for having taken up his challenge which Pasteur had issued at the January 29th meeting of the French Academy of Sciences. He then strongly requested the Academy to select a group of judges to evaluate the controversy between Bastian and himself.[9]

It was almost summer before definite plans for a meeting between Bastian and Pasteur in Paris had been arranged. Bastian was very eager to have the conditions of his demonstration before Pasteur and the Commission of the French Academy of Sciences thoroughly understood. He emphasized that the question under discussion between Pasteur and himself was *"whether previously boiled urine, protected from contamination, can or cannot be made to ferment and swarm with certain organisms by the addition of some quantity of liquor potassae which has been heated to 110° C., for twenty minutes at least.*" He also insisted that the board of judges confine its evaluation to the above question. Bastian did not want the Commission to interpret the issue under discussion with reference to either microbial spontaneous generation or to the germ theory of disease.[10]

Bastian left London on Friday, July 13, 1877, for his visit to Paris. On Sunday afternoon (the 15th), Bastian met with two members of the Commission in order to discuss plans for his meeting with Louis Pasteur. The men with whom Bastian spoke were Jean Baptiste André Dumas (1800-1884) and Henri Milne-Edwards (1800-1885). He told these two representatives of the Commission that his visit to Paris must be brief and that he did not have time to pursue alterations in the experiments, which the Commission might suggest, in the debate between Pasteur and himself. Upon learning this, Milne-Edwards flatly refused to serve on an Academy Commission which would be restricted from altering the experimental conditions under consideration. Bastian retorted that he had come to Paris to meet with the Commission in good faith and for a very specific reason, namely "to repeat certain well-defined experiments before them, and that they were commissioned to express an opinion thereon and on the experiments of M. Pasteur to the Academy of Sciences." Although he was unable to reach a compromise with Milne-Edwards, Bastian talked with Louis Pasteur and agreed to meet with the French scientist in Pasteur's laboratory on Monday morning, July 16th, in order to discuss experimental procedures for their proposed contest.

On Monday afternoon Bastian again met with Dumas. He offered a compromise arrangement to Dumas whereby during his first visit to Paris the Academy should restrict itself to reviewing the issue between Pasteur and himself as outlined in Dumas' April 25th letter to Bastian. Thereafter Bastian would return to London and await the Commission's findings on the first stage of the experiments. If the French Academy's Commission then wished to alter the experimental procedures, Bastian would travel again to Paris and there carry out the new investigations with Pasteur.[11]

Bastian and Pasteur were at their respective stations in Pasteur's laboratory at eight o'clock on Wednesday morning, July 18, 1877. Philippe Edouard Léon Van Tieghem (1839-1914), the third member of the Commission,[12] was also there. He was soon joined by Milne-Edwards who, having learned of the compromise between Bastian and Dumas and disapproving of that arrangement, left Pasteur's laboratory. For about an hour Van Tieghem also was away, but he returned to report that Milne-Edwards had departed because Dumas had not arrived. For another hour Bastian and Van Tieghem talked in a room above Pasteur's laboratory. In the interim Dumas had come to Pasteur's laboratory and, informed of Milne-Edwards' departure, he told Pasteur that the Commission was dissolved. Van Tieghem and Bastian only learned indirectly of Dumas' decision from Pasteur who was the only person remaining in the laboratory when they came downstairs. Since the members of the Commission failed to agree upon the conditions under which the experiments should be conducted, the contest between Bastian and Pasteur never did occur.[13]

John Tyndall, commenting on this proposed contest, expressed some concern for Pasteur's position in the debate. Tyndall's remarks were written more than seven years after the encounter between Bastian and Pasteur. It is not altogether clear whether Tyndall had the issues of the controversy clearly in mind. Certainly Tyndall's words do not apply to Pasteur's position on the treatment required to sterilize potash solution which Pasteur had set at 110° C. for twenty minutes. If his statement refers to the assumption made by both Bastian and Pasteur that five minutes' boiling at 100° C. was sufficient to sterilize acidic urine vessels in all instances, then perhaps Tyndall's remarks are most significant. Tyndall had previously experienced difficulties with bacterial endospores. Writing in December 1884, he observed that a statement in René Vallery-Radot's biography of Pasteur needed qualification. He pointed that Radot's sentence: "The great interest of Pasteur's method consists in its proving unanswerably that the origin of life in infusions which have been heated to the boiling point is solely due to the solid particles suspended in the air" was not literally true. Although many microbes are killed by boiling at 212° F. [100°C.], Tyndall emphasized that some germs, notably hay-bacillus endospores, can survive high-temperature treatments as severe as eight hours of boiling. His own experience with infusions prepared from desiccated hay caused Tyndall to be uneasy about the contest which was to have occurred in Paris between Bastian and Pasteur in July, 1877. He stressed that his concern was over the lethal temperature for microbes and not that spontaneous generation in the larger sense would have been vindicated.[14]

Had the encounter between Bastian and Pasteur actually taken place, perhaps both the heterogenist and the germ theorist might have claimed at least a partial victory. Microbes might have appeared in some of Bastian's boiled and subsequently neutralized urine flasks. On the other hand, Pasteur's careful procedures might have insured the sterility of his urine vessels especially if his flasks had been free of very desiccated bacterial endospores. What we do know is that the British heterogenist made an impact upon microbiology in France.

The best example of Bastian's stimulating effect upon Pasteur and his students is found in Charles Edouard Chamberland's (1851-1908) doctoral research project devoted to the study of microbial thermal resistance. Chamberland discovered that temperatures of at least 115°C. were required to destroy with certainty bacterial endospores if heat sterilization were to be achieved by other than prolonged temperature treatment such as boiling at 100°C.

for several hours. He also emphasized that glassware must be totally free of all germs before use in microbial investigations.[15] Commenting on Pasteur's and Chamberland's repetition of Bastian's experiments, the British bacteriologist and historian of bacteriology William Bulloch asserted that the sterilization of liquids began to be routinely carried out at 115 to 120° C. as a result of their investigations.[16]

The Urine Episode was another phase of the debate over microbial spontaneous generation in the 1870's. Studies on neutralized urine made microbial workers appreciate more clearly the high degree of heat resistance possessed by some bacterial endospores. Microbiologists also grasped the necessity for sterile glassware, sterile media, and sterile additives to these media if microbial experiments were to yield valid results.

Chapter 6

Factors Determining the Thermal Resistance of Infusions

Studies on turnip-cheese infusion, hay infusion, and urine during the 1870's enabled microbiologists to gain valuable insights into the maze of factors which determine the outcome of any heat-sterilization procedure. The results of these experiments provided important information on: (1) Circumstances of the Sterilization Process, (2) Characteristics of the Media, and (3)The Role of Different Kinds of Microbes.

Circumstances of the Sterilization Process

Investigators such as Thomas Henry Huxley, William Roberts, and John Tyndall had looked toward possible errors of technique in their initial attempts to account for the frequent failure of five minutes' boiling to sterilize organic solutions. A good example is found in Tyndall's reply to queries concerning the discrepancy between his own results and those obtained by Bastian and Burdon-Sanderson. Stressing the possibility of error in the experimental procedures of these researchers, Tyndall suggested the use of an oil bath in place of a Bunsen burner for boiling and a simple spirit lamp instead of a blow pipe for the sealing process.[1]

Tyndall's prolonged adherence to his position that the failure of five minutes' boiling as a heat-sterilization technique could be explained by technical errors led to an ironical situation. Tyndall, a leading proponent of the germ theory in England throughout the 1870's, unwittingly found himself lending support to the arguments of the great heterogenist Bastian who wrote sarcastically that Professor Tyndall "agrees with me in thinking that Bacteria and their germs are decidedly killed by five minutes' boiling in organic infusions. He still further supports the view held by me, in opposition to that of M. Pasteur, that such a result follows both with alkaline and with acid infusions."[2]

Beginning in autumn, 1876, Tyndall began to modify his position with regard to the first part of the above statement, and early the next year, his opinion concerning the second part of the statement had also been fully reversed. Nevertheless, Tyndall's realization that other factors in addition to experimental errors were involved in fully explaining failures of heat sterilization did not cause the British physicist to minimize the important ingredient of careful experimental technique. His comments on hermetic sealing in spring, 1877, indicate that such considerations remained very prominent in his mind.

Tyndall cautioned that hermetic sealing toward the end of boiling could be successfully accomplished only after considerable practice. He stressed the danger of introducing atmospheric germs during the sealing procedure when air pressure may vary considerably between the inside and outside of the flask. Thus vessels regarded as free of air might easily have been contaminated during the sealing process. Even a skilled technician, Tyndall asserted, might expect as high as 10% of his infusion vessels to be contaminated through defective sealing.[3]

Concepts of *thermal death point* and *thermal death time* were also developed during the 1870's. In their studies of turnip-cheese infusions, John Burdon-Sanderson and Edwin Ray Lankester had focused upon the relationship between intensity of temperature and time of exposure to a given temperature during sterilization. Lankester had described a curve indicating the rate of sterilization during heat treatment by plotting the sterilization temperature against the time of exposure to that temperature.

Thermal Resistance of Infusions 67

The two factors of time and temperature were also stressed by William Roberts in spring, 1874, when he pointed out the correlation between intensity of temperature and duration of heating during the sterilization process. Roberts attempted to assign equivalent time and temperature values to different methods of heat sterilization. His famous paper of that year remarked:

> There appeared to be two factors of equal importance in the process of sterilization, namely, the *degree* of heat and the *duration* of its application. These two factors were mutually compensatory, in such fashion that a longer exposure to a lower temperature was equivalent to a shorter exposure to a higher temperature. For example, speaking roughly, an exposure for an hour to a heat of 100° Cent. was equivalent to an exposure for fifteen minutes to a heat of 109° Cent.[4]

To some extent the great concern for experimental errors expressed by germ theorists such as Huxley, Roberts, and Tyndall (from 1870 to autumn, 1876) was justified. Even today heat-sterilization procedures are known to fail occasionally if careless technique and haste are permitted to creep into the operation. Thus present-day bacteriologists may on occasion find spore-forming bacilli in supposedly pure cultures of other bacteria or in culture media thought to have been sterile.[5]

A present-day authority on sterilization techniques stresses the variability of heat resistance caused by different cultural factors. He suggests that thermal death time is a far more meaningful criterion of heat resistance than is the expression thermal death point.[6] The heat sensitivity of a particular microbe is measured in terms of the thermal death point and thermal death time assigned to that microorganism. Thermal death point refers to the temperature at which a given microorganism is destroyed by the action of heat. Associated with a specific thermal death point is a definite thermal death time which refers to the duration of heating required to kill a microorganism at a specific temperature.

Very frequently the thermal death point of a specific bacterium is given for a thermal death time of ten minutes.[7]

The application of 90°C. for ten minutes to an organic solution is lethal to all forms of microbial life except for bacterial endospores.[8] Efforts to carry out sterilization procedures in a laboratory environment which is excessively contaminated with bacterial endospores, can, even today, present many of the frustrations which John Tyndall experienced in his laboratory at the Royal Institution in winter 1876-77. Furthermore, the surrounding circumstances with regard to the oxygen supply are of great importance in heat-sterilization procedures which depend upon the germination of endospores of aerobic as well as anaerobic bacteria. Such a process is John Tyndall's discontinuous heating to be discussed below (Chapter 7).

Characteristics of the Media

By the year 1874 it was well known that the physical and chemical properties of substances to be rendered free of microbes definitely influenced the ease or difficulty of heat-sterilization procedures. William Roberts had observed that:

> Organic liquids and mixtures are capable of being permanently sterilized by the heat of boiling water. Some are sterilized by an exposure to this heat for a few minutes, others require an exposure of twenty to forty minutes, and others an exposure of one, two, or several hours. A slight difference in their reaction was sufficient to alter very greatly the amount of heat required for their sterilization.[9]

When referring to the "aggregation of the materials" Roberts meant the influence that the presence of solid particles may exert upon heat-sterilization efforts as opposed to the treatment of purely liquid organic solutions. In his examination of turnip-cheese infusion, Edwin Ray Lankester concentrated on the protective nature of the cheese particles present in an otherwise purely liquid turnip infusion.

Thermal Resistance of Infusions

Lankester believed that germs enclosed by the cheese particles were effectively protected from the action of heat and moisture. Likewise John Tyndall, a latecomer to turnip-cheese studies, expressed his conviction that cheese and its insulating and impervious qualities were responsible for the well-known difficulty of sterilizing turnip-cheese infusion. As a result of his investigations conducted on turnip-cheese infusion during summer, 1876, Tyndall discussed the phenomenon of solid particles shielding microbes from the destructive action of high temperatures.

> Not temperature alone, but the ability to diffuse its juices or salts, appears to be a condition of importance in the destruction of the integrity and life of a germ by boiling water. Without diffusion a germ may withstand temperatures competent to destroy it where diffusion is free. I need not remark on the imperviousness of cheese to water, and its consequent power to prevent diffusion.[10]

Although Louis Pasteur had noted back in 1862 that milk required temperatures of 105° to 110°C. to free it of microbes, it was not until late in the 1870's that the phenomenon of alkaline protection as opposed to acidic predisposition to the detrimental effect of high-temperature treatments was generally recognized and accepted. John Scott Burdon-Sanderson had stressed that he used neutral turnip-cheese infusions in all of his experiments on that medium. William Roberts had pointed out the surprising resistance of neutralized or slightly alkaline hay infusion to the action of elevated temperatures. Until early 1877, however, John Tyndall had denied the validity of William Roberts' observation of the protective effect of an alkaline reaction. Through studies on urine, it gradually became clear for all times that differences in pH reaction do indeed profoundly affect the outcome of a given heat-sterilization effort.[11]

Chapter 6

Especially interesting during the 1870's were the theories offered to explain the protective action of alkaline media for the germs which they contained. For example, Pasteur observed about the surprising heat resistance of neutral or slightly alkaline urine: "The facts do not prove spontaneous generation, but only that certain germs resist a temperature of $100°$ [C.] in neutral or slightly alkaline media, their envelopes, doubtless, not being penetrated in this case by the water."[12] And the following year Tyndall commented on bacteria, fungi and alkaline and acidic media: he remarked that *Penicillium* grows readily in acidic infusions whereas bacteria generally are found in neutral or alkaline media. He disagreed with Pasteur on the mechanism by which neutral or slightly alkaline organic solutions resist sterilization:

> The best thought that I have been able to bestow upon this subject does not induce me to lean towards the explanation suggested by M. Pasteur, namely, that the germs escape the destructive action of the heat because they are not wetted by the alkaline or neutral liquid. From the comparative action of alkalized and acidulated water upon hay, I should be inclined to infer that the wetting of its germs by the former would be more prompt than by the latter. The question, I think, is not one of wetting, but of relative nutritive power.

Tyndall also explained in detail his own concept of greater bacterial thermal resistance in alkaline rather than acidic media. Emphasizing the importance of the pH reaction of culture media, he observed that "*Bacteria*-germs are more fully vivified and better nourished in neutral than in acid ones" and that "a germ brought close to the death-point in a neutral or an alkaline infusion may revive, while in an acid one it may perish — just as proper nutriment may rescue a dying man while improper nutriment would fail to do so."[13] Writing in 1874, Roberts offered some perceptive comments on the protection afforded to germs by alkaline solutions. His theory, very similar to the modern view, held that the heat resistance of alkaline liquids "was probably due to the fact that albuminoid matter is more easily and more completely coagulated by a given heat in acid than in alkaline solutions."[14]

Current appraisals of heat sterilization remind us of Edwin Ray Lankester's concern for "through heating" during his experiments in the turnip-cheese episode. The presence of solid particles can make a great deal of difference in the outcome of a heat-sterilization process even with present-day procedures. Modern guides to sterilization procedures insist that the very center of the material must be subjected to the destructive temperature for a sufficient time. This requirement is indispensable if the sterilization process is to succeed. Containers of meats or vegetables must be subjected to autoclave treatment for 1½ hours to 2 hours in order to make certain that any bacterial endospores at the center of these materials have been destroyed.[15] Another microbiologist of today in discussing factors influencing microbial thermal resistance indicates that colloids tend to protect microbes from rapid destruction by heat sterilization. He notes that lactic acid bacteria are killed at a much lower temperature in peptone water than in media such as milk or cream.[16]

Today we know that bacteria usually grow in the range pH 4 to 9. Although yeast and fungi grow well under acidic conditions, most bacteria will not grow at a pH below 4.[17] Whereas mould and yeast heat resistance occurs under acidic conditions, bacterial endospores are most resistant at pH 6 to 8. In the case of *Bacillus subtilis* resistance drops off sharply at pH 6.6. In fact, the resistance of most organisms appears to fall off sharply at and just below a specific pH value.[18] As Roberts astutely suggested in the 1870's, acidic conditions may enhance the destructive power of heat-sterilization procedures by hastening the coagulation of protein. Likewise protein complexes such as enzymes are equally or even more sensitive to pH changes as to temperature alterations. Enzymes have definite limits of tolerance to pH levels as well as to temperature values. In this respect enzymes and living cells possess similar responses.[19]

The next category of factors which determines the efficacy of any heat-sterilization effort includes characters of the microorganisms present in a given organic solution when it is subjected to a heat-sterilization process.

Role of Different Kinds of Microbes

There was a gradual transition from concern over possible errors to explain the failure of 100°C. to sterilize all organic solutions to consideration of the overall circumstances of a given heat-sterilization process. This trend led gradually to the realization that properties of the medium to be sterilized also exert a profound effect upon the outcome of any effort to do so. The further movement from concentration upon the properties of the medium to concern for the characteristics of the microorganisms themselves was suggested in Edwin Ray Lankester's comments during the turnip-cheese episode. He believed that more knowledge of the life histories of bacteria might explain many of the puzzling aspects of microbial sterilization. William Roberts commented on this matter and emphasized the heat resistance of bacteria in contrast to the heat susceptibility of the fungi. In his excellent paper of 1874 Roberts wrote:

> No organic liquid or mixture, subjected for however short a time to the heat of boiling water, ever produced (provided there was no fresh infection) any fungoid or torulaceous [yeast-like] organisms. If germination took place, the organisms produced invariably belonged to the great group of *Bacteria.* This appears to indicate that *Torulae* and their germs are more easily destroyed by heat than *Bacteria* and their germs.

Roberts continued by noting variable heat resistance within the bacteria themselves. He suggested that "it does not seem unreasonable to suppose that different races of *Bacteria*, or different phases of their development, are capable of offering very different degrees of resistance to the destructive influences of heat."[20] His insights are vindicated today, when we know that the genus *Bacillus* and the genus

Clostridium contain bacteria that form endospores which are extremely heat resistant. These bacterial endospores possess far greater thermal durability than the conidiospores, sporangiospores, ascospores, and conidia of other fungi. No other spores approach the thermal resistance of bacterial endospores.[21]

In his studies of turnip-cheese infusion, Ferdinand Cohn discovered endospore formation in the butyric bacillus, a bacterium now termed *Clostridium butyricum*. This motile bacterium is anaerobic (killed by the presence of free oxygen), a property first observed by Louis Pasteur in 1861. Another distinguishing characteristic of *Clostridium butyricum* is its manner of spore germination. When the new bacillus or rod-like vegetative cell develops from a butyric-bacillus spore it emerges from one end of the spore.[22]

Investigations on hay infusion led to Ferdinand Cohn's study of the life history of the hay-bacillus, *Bacillus subtilis*. Like the butyric bacillus, the hay bacillus is a motile, rod-like bacterium. However, *Bacillus subtilis* is aerobic in its oxygen relationships — that is, it requires free oxygen for its life processes. Another distinguishing feature between these two organisms is their distinctly different manner of spore germination. Unlike the "polar emergence" characteristic of *Clostridium butyricum*, spores of *Bacillus subtilis* split at the center of the spore and the new hay-bacillus cell emerges equatorially.[23]

Both the butyric bacillus and the hay bacillus reproduce by fission — that is, by division of the vegetative cell. The endospores of these microbes are *not* a means of reproduction; rather, their endospores provide a mechanism for survival during adverse periods. In both these bacteria a simple vegetative cell, under conditions not entirely understood, may form a durable resting body within the confines of the vegetative cell (hence the term *endospore*). With the return of a favorable environment with regard to moisture, oxygen, nutrition, and pH reaction a given endospore may germinate and produce a single vegetative cell or bacillus. Thus in the two organisms we have studied, an individual vegetative cell gives rise to one endospore which in turn may germinate to form a new bacillus microbe.[24]

A number of factors influence the durability of even these highly-resistant bacterial endospores. Tyndall's perceptive comments regarding "germs in the air" certainly apply to bacterial endospores.

> There cannot, I think, be a doubt that the germs in the air differ widely among themselves as regards *preparedness* for development. Some are fresh, others old; some are dry, others moist. Infected by such germs, the same infusion would require different lengths of time to develope Bacterial life.[25]

Hay-infusion studies revealed the amazing heat resistance of bacterial endospores, which can withstand the temperature of boiling water for several hours. Such instances of extraordinary heat resistance Tyndall attributed primarily to the extreme dryness of the spores which enabled them to resist the action of moisture and high temperature. Tyndall's suggestion of low moisture content is still mentioned in modern discussions of microbial thermal resistance.[26] Current views regarding the mechanism of microbial durability, however, are far more complex than Tyndall's analysis. Although a truly satisfactory explanation of the surprising heat resistance of bacterial endospores has not yet been formulated, there are several suggested reasons for endospore durability.

We should understand the mechanism by which heat does manage to kill these durable resting bodies. One authority states that heat inactivates proteins and enzymes essential to proper microbial cell functions, thus killing the microorganism.[27] This statement immediately suggests to us that bacterial endospores must possess particularly resistant enzymes and proteins. Such, indeed, is believed to be the case.[28]

Another contemporary microbiologist stresses that spores and heat-tolerant bacteria possess heat-resistant enzymes. Thermophilic bacteria may grow in the 70 — 80°C. temperature range. The same investigator points out that some enzymes are simply quite heat resistant. Thus the thermostability of spores may be brought about by dehydration, influence of calcium dipicolinate, peptone and fat protective shielding, and intrinsic resistance.[29]

Other accounts frankly state that the extraordinary heat resistance of bacterial endospores is still a mystery. Compared to vegetative bacterial cells, endospores contain less water. Lower water content may make proteins in the endospores less subject to denaturation when heat is applied. Furthermore, in endospores protein substances tend to be in an insoluble form and are often associated with lipids. It is known that one insoluble enzyme in these spores, alamine racemase, is not easily denatured by heat. If this enzyme is removed from the spores and treated chemically to make it soluble, however, then the enzyme becomes as liable to heat destruction as the same enzyme contained in vegetative cells.[30]

In summary, many factors govern the rate of death of a particular kind of bacterium in a given material. Such factors include: acidity and moisture, temperature, type of bacterial cell, age of the cells, and the number of bacteria.[31]

Chapter 7
Heat-Sterilization Procedures of the 1870's

There is a reason for restricting our discussion of sterilization methods during the 1870's solely to techniques of heat sterilization. The justification for our narrow focus stems from the experience of practicing microbiologists. For example, in 1884 George M. Sternberg expressed his high regard for heat-sterilization procedures. He stated that the best technique was to place a fluid to be sterilized in a germ-proof container and then to subject it to a temperature sure to kill any microorganisms present in the fluid.[1] Today we also consider the application of heat by one means or another to be one of the most efficient methods for destroying the bacteria,[2] and today it is generally agreed that heat sterilization by means of moist heat subjected to pressure is the most dependable method of sterilization.[3]

Both moist and dry methods of heat sterilization were employed throughout the 1870's. Moist heat kills microorganisms by causing coagulation of proteins and enzymes; dry heat destroys microbes through oxidation or by actual burning.[4] During the 1870's methods of sterilization by moist heat were applied to organic solutions by: Prolonged Boiling at 100°C., Heat-Treatment above 100°C., and Discontinuous Heating or Tyndallization. After examining these techniques for sterilizing infusions by moist heat, the Dry-Oven Method of Heat Sterilization will be discussed.

Prolonged Boiling at 100° C.

Although boiling at 100°C. for five or ten minutes to render organic solutions free of microbial life was widespread in the 1870's, gradually this heat-sterilization procedure came to be discredited as a reliable laboratory technique in microbiology. The realization that five or ten minutes' boiling frequently did not suffice to destroy microorganisms cleared the way for more effective methods of heat sterilization. Nevertheless, efforts to sterilize infusions simply by boiling them for short periods of time at 100°C. persisted until the middle of the 1870's. John Tyndall, for example, maintained until 18 January 1877 that "five minutes' boiling" was an effective heat-sterilization technique for purely-liquid infusions.[5]

Microbe investigators became dissatisfied not only with the frequent failure of sterilization attempts by brief periods of boiling, but they also complained of the problems inherent in simple boiling techniques. In 1874 Roberts wrote of his dissatisfaction with the heat-sterilization method using a flame to boil the contents of a plugged flask directly. He noted that his materials underwent a change of concentration when they were boiled for 15 to 20 minutes. Furthermore, Roberts pointed out that temperature grades fluctuated depending upon the tightness of a given flask's plug. He remarked that the boiling-point was 3° to 6°C. above the usual temperature in his plugged flasks.[6]

Procedures in which infusion vessels were "boiled briskly over the flame" were replaced by the technique of submerging hermetically-sealed infusion vessels "in a can full of water. The can was next placed over a source of heat and boiled for the required time."[7] In 1873 the use of these water baths in an attempt to sterilize milk was described by Joseph Lister, who used a saucepan of boiling water to heat immersed flasks of milk for an hour. In this procedure Lister made certain that the boiling water of the saucepan never fell below the level of the milk contained in the flasks. He noted that frothing of the milk so common in direct-flame boiling was eliminated by the water-bath technique which had been suggested to him by Mr. Godlee of University College, London. Furthermore, sputtering previously described by Dr. Roberts of Manchester was avoided, as was water loss through evaporation of the material in the flasks. Thus the specific gravity of the organic matter remained unchanged. This eliminated the danger of conditions which might not permit organic development. Finally, the milk was not altered chemically by this sterilization technique.[8]

Emphasizing that "sterilization" means killing all microorganisms, including the most durable bacterial endospores, one modern authority on sterilization procedures cautions that there is a frequent misconception among laymen that simply boiling microorganisms kills them with certainty. Nothing could be further from the truth.[9]

Heat-Treatment above 100° C.

Besides attempts to sterilize infusions by simple boiling at 100° C. for five or ten minutes (a procedure which is effective only in the case of purely liquid infusions which do *not* contain *Bacillus* or *Clostridium* endospores) or by the use of water baths (effective for all infusions if boiling is continued for many hours), there were two other widespread practices of heat sterilization during the 1870's. Both of these methods required that infusions be subjected to temperatures above 100° C. These procedures made use of pressure-cooker-like devices such as the Papin digester and baths of boiling oil, paraffin, or brine solution.

John Scott Burdon-Sanderson employed the Papin digester[10] to boil his turnip-cheese infusions at pressures greater than atmospheric pressure and hence at temperatures above 100° C. He boiled and sealed his flasks and "placed them in a digester in which they were subjected to ebullition under a pressure of two inches or more of mercury." No bacteria subsequently appeared in the fluids treated by Burdon-Sanderson.[11]

Baths composed of a liquid which had a boiling point higher than the temperature of boiling water comprised another method of subjecting infusions to temperatures above 100°C. Working with hermetically sealed infusion vessels, William Roberts used these high temperature baths. Besides preventing the evaporation of his organic solutions, Roberts remarked that a further advantage of oil or brine baths was that they enabled the microbe investigator to make comparative studies of the effect of different time and temperature treatments upon infusions. For example, he recorded that "an exposure for an hour to a heat of 100°Cent. was equivalent to an exposure for fifteen minutes to a heat of 109°Cent."[12]

George M. Sternberg left an interesting account of the use of these high temperature baths in America. In summer 1881 Sternberg was engaged in a research project at Johns Hopkins. He found that the biology laboratory there was infected with bacillus spores and that one species of bacteria formed little islands on the surface of his culture media. This occurred even after the cultivating fluid had been boiled for over an hour. Sternberg thereupon used baths of paraffin or brine to raise his hermetically sealed culture flasks to 105°C. for ½ to 1 hour. He found that the bacillus spores were effectively destroyed in his cultivating fluids.[13]

In 1885 Charles Sumner Dolley wrote about the use of high temperature baths as a heat sterilization method. He described as "Pasteur's Sterilization Method" the technique of sealing cultivating materials into fused glass tubes previously blown into small flasks or balloons. A chloride of lime or nitrate of soda bath surrounded vessels held in place by a perforated metal plate at 110°C. to 115°C. for 12 hours.[14]

From Pasteur's observation that a calcium chloride bath of over 100°C. was effective for sterilizing hermetically sealed flasks, heat sterilization procedures evolved to the use of superheated steam. Flasks to be sterilized were simply provided with wool plugs and then placed in a container resembling in principle the digester for softening bones devised in 1681 by Denys Papin. By 1884 the Weisnegg engineering company in Paris had produced early commercial versions of Chamberland's autoclave. These early laboratory models of the autoclave resembled their more sophisticated modern counterparts.[15]

A home pressure cooker is essentially an elementary autoclave. Regardless of the size of an autoclave or its complements of wheels, valves, pipes, gauges and clocks, the apparatus has as its purpose the sterilization of materials through the agency of heat applied as steam under considerable pressure.[16] Even with modern autoclaves a surprising amount of time may be required to effect sterilization. For example, flat sours in canned foods are caused by bacterial spores which are extraordinarily heat resistant. Such microbes can only be destroyed with certainty by being subjected for three hours to steam under pressure at 120°C. However, most bacterial spores are killed within 15 to 20 minutes by moist heat at 115 to 120°C.[17]

One present-day authority on sterilization procedures related a personal experience concerning microbial resistance to the autoclave. Years ago George Sykes found that an unidentified microorganism in meat kept appearing in his culture media. The microbe, or rather its heat-resistant spores, withstood flowing steam for one hour followed by 45 minutes of steam pressure at 115°C. The organism was finally killed when steam pressure at 115°C. was applied for 90 minutes or at 121°C. for 60 minutes.[18] In the 1870's, microbial sterilization was also achieved by the use of moist-heat treatments below 100°C.

Discontinuous Heating or Tyndallization

A third method of heat sterilization employed during the 1870's and indeed still performed in microbiology laboratories today is John Tyndall's method of *discontinuous heating*, also referred to as Tyndallization. This heat-sterilization technique was developed as a result of Tyndall's experiences in the study of hay infusion.

Tyndall had grasped the significance of heat sensitivity in the vegetative cells of *Bacillus subtilis* as opposed to the high degree of heat resistance exhibited by hay-bacillus endospores. He described his method of discontinuous heating in his letter of 14 February 1877 to T.H. Huxley who was then Secretary of the Royal Society of London.[19] In that letter Tyndall was careful to emphasize four essential points. He stated that by "following up the plain suggestions of the germ theory" he had achieved sterilization of all organic solutions "even in the midst of a virulently infective atmosphere" by heat treatment at a temperature below 100° C.

Although it is true that Tyndall himself often used 100° C. for very brief periods to effect discontinuous heating, it by no means is a requirement of the process. Subsequent accounts of Tyndallization frequently mislead the reader by conveying the impression that Tyndall's technique of discontinuous heating requires the temperature of 100° C. in order to sterilize infusions.[20] This is *not* essential for the success of Tyndall's method.

Tyndall also presented the apparent paradox of his method in terms of contemporary views regarding temperature and time relationships in heat-sterilization procedures.

> I boil an infusion for fifteen minutes, expose it to a temperature of 90°Fahr. [32.2°C.], and find it twenty-four hours afterwards swarming with life. I submit a second sample of the same infusion to a temperature lower than that of boiling water for five minutes, and it is rendered permanently barren.

Next, Tyndall described his actual technique of sterilization accomplished by applying a surprisingly low temperature treatment to infusions:

Heat-Sterilization Procedures

> This then is my mode of proceeding: —Before the latent period of any of the germs has been completed (say a few hours after the preparation of the infusion), I subject it for a brief interval to a temperature which may be under that of boiling water. Such softened and vivified germs as are on the point of passing into active life are thereby killed; others not yet softened remain intact. I repeat this process well within the interval necessary for the most advanced of those others to finish their period of latency. The number of undestroyed germs is further diminished by this second heating. After a number of repetitions, which varies with the character of the germs, the infusion, however obstinate, is completely sterilized.

Expressed in present-day terminology Tyndall might have explained that his application of temperature treatment lower than the boiling point of water came after the germination of some bacterial endospores and thus killed any vegetative cells which had developed within a few hours after he had prepared his infusions. By allowing many hours to elapse in the intervals between these brief and rather mild heatings (over 60° C., but well under 100° C.), Tyndall permitted other endospores to soften and to develop into new vegetative cells which in turn were destroyed by the next heating. Repeating this procedure four, five, or even six times generally sufficed to free Tyndall's infusions of microorganisms.

The final significant comment in his letter to Huxley written in early 1877 compared the duration of heating in more conventional heat-sterilization techniques to that in the method employed by Tyndall himself.

> The periods of heating need not exceed a fraction of a minute in duration. Sum them up in the case of an infusion which they have perfectly sterilized; they amount altogether to, say, five minutes. Boil another sample of the same infusion continuously for fifteen or even sixty minutes, you fail to sterilize it, although the temperature is higher and its time of application more than tenfold that which, discontinuously applied, infallibly produces barrenness.

In May 1877 Tyndall wrote of his use of 100°C. to effect sterilization by discontinuous heating. He had constructed closed chambers or box-like structures designed to shield exposed, but sterile infusions from subsequent contamination by air-borne microorganisms or microbial spores. To destroy the microbes in his open vessels contained by his chambers, Tyndall carried out his discontinuous heating procedure. He initially submerged his test tubes in a 300°F. [148.9°C.] oil bath. After a (30 seconds or less) boiling, Tyndall removed the oil bath. He also sometimes exposed his test tubes to boiling water for 2 to 3 minutes and then continued to heat them with a spirit lamp. In his opinion the spirit lamp technique provided greater control to the experimenter than did the oil-bath procedure. Every 12 hours Tyndall repeated his heating process except that the time interval was shortened when especially nutritive infusions had been placed in a warm area. He cautioned that reheatings must be carried out before visible changes occurred in the infusion, and he stressed the value of experience in performing this method of heat sterilization.[21]

Tyndall claimed that "with due care the method of sterilization here described is infallible, however highly infective the surrounding atmosphere may be." We now know that quite the contrary is true of discontinuous heating. Tyndallization is not an effective method of sterilization unless the material under treatment favors spore germination. Many bacterial spores will not germinate in non-nutrient solutions. Since the heat-sensitive vegetative phase of the microbe is the portion of its life history which is vulnerable to the discontinuous principle, the spores which do not germinate will always present a threat of future contamination.[22]

Especially in water bacterial spores may not germinate. Anaerobic spores in the presence of free oxygen may also fail to germinate and thus remain as possible sources of contamination at future stages of experimentation. Similarly, if aerobic spores are deprived of air they will not germinate satisfactorily and may present problems later on.[23] Despite these shortcomings, Tyndallization is used even today for destroying microbial life in egg and serum media and in other heat-sensitive cultivating materials.[24]

Dry-Oven Method of Heat Sterilization

The major differences between moist heat and dry heat in sterilizing materials include their mode of action, the techniques by which these agencies of microbe destruction are applied, and the length of time required for the accomplishment of this purpose. Moist heat kills microbes by means of coagulation of proteins and enzymes; dry heat destroys microorganisms by oxidation or by actual burning. Moist-heat techniques of sterilization generally require more complex equipment than do dry-heat procedures. However, methods of dry-heat sterilization require rather long periods of time in order to be effective.

Burdon-Sanderson had suggested in December, 1872, to Bastian that the latter insure the sterility of the interior of his glass vessels by subjecting his flasks to the temperature of 250° C. prior to their use in Bastian's third set of experiments.[25] In 1874 William Roberts also described his use of dry heat to sterilize glassware about to be employed in his experiments. He sterilized his tubes by flaming them with a spirit lamp until their cotton-wool plugs began to char.[26]

While mentioning these attempts to accomplish dry-heat sterilization by direct flaming, it is interesting to note that flaming of glassware has not received a very favorable evaluation from one modern authority of microbial sterilization techniques. That investigator has pointed out that the frequent use of flame sterilization probably makes laboratory workers feel better psychologically. Although the practice may be of value to kill microbes present on platinum loops and cotton wool plugs, he expresses doubt that the flaming of vessel necks accomplishes very much.[27]

In 1878 Joseph Lister described his use of a special oven to effect dry-heat sterilization. Lister found that his glassware could be sterilized by 2 hours' exposure to 300° F. (148.9° C.). He made sure that air entering his vessels during cooling was filtered and thus free of dust and its associated germs. He therefore used a cast-iron box for his dry-heat sterilization oven. Using an abundant supply of cotton wool, Lister tightly sealed the oven door. The cotton wool filtered the air which entered the oven during cooling. In this way he avoided the danger of permitting the heat physically to destroy the cotton filter. His cotton was browned uniformly at the bottom as well as at the top of the oven. Lister accomplished even distribution of heat by having the heat from a large Bunsen burner shielded from his oven by three shelves of sheet iron. The heat from the Bunsen burner was in fact radiated over the entire exterior of Lister's dry-heat oven. The thermometer of the stove indicated that the temperature was 300° F. (148.9° C.) to 350° F. (176.8° C.) for two hours. After cooling, Lister removed his covered glassware from the oven, and he felt confident that all microorganisms had been destroyed.[28]

There is lack of agreement regarding essential time and temperature values sufficient for dry-oven methods of heat sterilization. For example, early in this century one microbe investigator found that plates, pipettes, culture dishes, test tubes, and flasks could be sterilized by exposure for one hour to a temperature of 150° to 180°C.[29] A present-day microbiologist recommends that baking at 165°C. for two hours be used to insure killing the most durable bacterial endospores which may be found on glassware.[30]

Commenting on the United States Pharmacopeia recommendation that dry-heat sterilization should be carried out at 170°C. for a period of two hours, a modern British researcher states that such a procedure seems excessive. Personal experience has convinced him that only a truly extraordinary microbe can withstand one hour's exposure to

150°C.[31] Some American microbiologists disagree. Since microbes are able to withstand dry heat more than steam they recommend hot-air sterilization at 170°C. for 1½ hours.[32]

Another recent and rather complete account of dry-heat sterilization points out that in oven sterilization marked coagulation is lacking because of the absence of moisture. Thus oven temperatures of 165° to 170°C. or 329° to 338°F. are required to kill spores on articles. Although there is no coagulation, slight charring of microbes occurs at 165°C. for two hours and this is considered to be an effective dry-heat sterilization treatment.[33]

The four major categories of heat sterilization employed during the 1870's still find certain kinds of use in the home or laboratory in our own day. For example, boiling is often used to render food and water, if not totally free from bacterial endospores, at least safe for consumption. Furthermore, heat-sterilization methods carried out at temperatures above 100°C. led to the development of procedures that were to culminate in the first commercially-produced autoclaves in Paris in 1884. Despite the many limitations of discontinuous heating, Tyndallization is used in the laboratory even today when especially heat-sensitive substances are to be sterilized. Heat-sterilization procedures, relying upon the dry-heat or dry-oven method of sterilization, still find wide application for sterilizing laboratory glassware to be used in microbial experiments.

By the close of the 1870's, there were very effective methods of heat sterilization available to microbiologists. Thus researches into microbiology could proceed without the former confusion of the early 1870's which frequently arose from impure cultures and laboratory contaminants.

Conclusion

Many problems plagued microbiology in the early 1870's. Most of the scientists working with microbes were educated in fields far removed from the detailed study of microorganisms. Ferdinand Cohn, trained as a professional botanist, was a notable exception. Investigators labored under technical shortcomings such as inadequate microscopes, primitive staining procedures, and lack of pure-culture techniques. Important gaps in microbial knowledge included ignorance of bacterial endospores and confusion over microbial specificity, pathogenicity, distribution, nutritional requirements, and oxygen relationships. The widespread use of ineffective heat sterilization procedures such as boiling organic solutions at 100° C. for five to ten minutes only added to the sometimes perplexing results obtained. Frequently this heat treatment failed to protect boiled and hermetically sealed infusion vessels from the subsequent appearance of microbes. Such failures were interpreted by Henry Charlton Bastian as instances of the origin of microorganisms from organic materials entirely independently of pre-existing germs.

Henry Charlton Bastian, John Burdon-Sanderson, Edwin Ray Lankester, Ferdinand Cohn, William Roberts, and John Tyndall were involved in the turnip-cheese episode, an important phase of the spontaneous-generation debate of the 1870's. Ferdinand Cohn's study of cheese manufacture, culminating in summer 1873 with his discovery of *Clostridium butyricum* endospores, elucidated British experiments on organic solutions. In this instance, durable endospores of the butyric bacillus accounted for Bastian's claims of the *de novo* origin of microbes.

British investigations on the response of hay infusion to high-temperature treatment eventually led to Ferdinand Cohn's discovery of hay-bacillus endospores. Cohn's paper on *Bacillus subtilis* exerted a profound influence on John Tyndall who admitted in early 1877 that five minutes' boiling was insufficient to sterilize all purely liquid organic solutions. Out of his studies on hay infusion, the British physicist devised his method of discontinuous heating to sterilize culture media. Here Tyndall's tendency to praise Robert Koch while ignoring Ferdinand Cohn is extremely puzzling.

Bastian performed experiments on urine in the 1870's to support his belief in heterogenesis. Both William Roberts and John Tyndall obtained results contradictory to Bastian's findings but corroborating Pasteur's earlier observation that organic solutions of neutral or slightly alkaline reaction are more difficult to sterilize than acidic infusions.

The spontaneous-generation debate of the 1870's produced information of great value to microbiologists. Very significant was an appreciation for the importance of the circumstances surrounding any heat sterilization process, of the characteristics of the media to be sterilized, and of the role played by different kinds of microbes in determining the thermal resistance of germs. Such information led to the realization that the overall circumstances of a heat sterilization process such as laboratory contamination or oxygen relationships may influence the outcome of a sterilization effort; experimental error was not to blame. The presence of solid particles, an alkaline pH reaction, and the contamination and desiccation of the medium likewise determine the ease or difficulty of a heat-sterilization effort. During the 1870's it was recognized that bacteria are more heat resistant than the fungi and that there are extraordinary examples of durability to high temperature and to extreme dryness among

Conclusion

the bacteria. Both the genus *Clostridium* and the genus *Bacillus* form bacterial endospores. *Clostridium butyricum* was responsible for the mysterious resistance of turnip-cheese infusion to high-temperature treatment. The hay bacillus, *Bacillus subtilis*, remains even today the subject of various modern theories to explain the unusual heat resistance of bacterial endospores.

A variety of heat sterilization techniques were used by microbial workers during the 1870's. Heat sterilization under moist conditions takes less time than under dry circumstances. Thus the procedure of boiling organic solutions at $100°$ C. for variable lengths of time was widely applied throughout the decade. Heat treatment above $100°$ C. was carried out during the same period with the Papin digester or by means of hermetically sealed vessels submerged in baths of boiling brine solution, oil, or paraffin. Early in 1877 John Tyndall devised his method of heat sterilization by discontinuous heating as a result of his experiences during the hay-infusion controversy. The technique of dry-oven heat sterilization is still used to sterilize laboratory glassware.

The dispute over microbial spontaneous generation throughout the 1870's stimulated microbiological progress during that period by leading to investigations into the biology of microorganisms, to an appreciation of the various factors affecting the thermal resistance of germs, and to the development of efficient and effective means of heat sterilization designed to cope with these factors. The breakthroughs characteristic of the new science of microbiology in the 1870's were to provide a foundation for the "Golden Age of Bacteriology" which began in the early 1880's with the microbiological techniques developed by Robert Koch and his students.

Footnotes

Chapter 1

1. *Report of the British Association for the Advancement of Science* (1870) Liverpool, pp. lxxv-lxxvi.
2. *Ibid.*, p. lxxxiii.
3. To appreciate the attention which Huxley's speech received and the impact which his words had upon the scientists of his day, see *Nature* for:
22 September 1870 vol. 2, p. 417
29 September 1870 vol. 2, p. 438
5 January 1871 vol. 3, p. 181
13 June 1872 vol. 6, p. 132
16 August 1877 vol. 16, pp. 303-304

Chapter 2

1. Wyville Thomson, "Fermentation and Putrefaction," *Nature*, 7 (1872), 79.
2. Ferdinand Julius Cohn, *Bacteria: The Smallest of Living Organisms* (1872), Translated by Charles Sumner Dolley (1881), (Baltimore, 1939), p. 15. Also see:
Ferdinand Julius Cohn, *Ueber Bacterien, Die Kleinsten Lebenden Wesen*, (Berlin, 1872), p. 6. Cohn's monograph may be found in: *Sammlung Gemeinverständlicher Wissenschaftlicher Vorträge*, 7th series containing Numbers 145-168 (1872-73), Number 165, 3-35.

Also see: Ferdinand Julius Cohn, "Untersuchungen über Bacterien I," *Beiträge zur Biologie der Pflanzen*, Bd. 1, Heft 2 (1872), 184. [Publication date for Heft 2 is given in *Heinsius Allgemeines Bücher-Lexikon*, Bd. 15 (1877), 316. Note that the title page for Band 1 of the *Beiträge* bears the date 1875.]

3. William Roberts, "Studies on Biogenesis," *Philosophical Transactions of the Royal Society of London*, 164 (1874), 472.

4. Benjamin Dawes, *A Hundred Years of Biology*, (London, 1952), p. 39.

5. Robert Koch, "Untersuchungen über Bacterien VI; Verfahren zur Untersuchung, zum Conserviren und Photographiren der Bacterien," *Beiträge zur Biologie der Pflanzen*, Bd. 2 Heft 3 (1877), 399-434.

6. William Bulloch, *The History of Bacteriology* (London, 1938), pp. 222-223. Also see: Roger Y. Stanier, Michael Doudoroff, and Edward A. Adelberg, *The Microbial World*, 2nd Ed. (Englewood Cliffs, N.J., 1963), pp. 33-34.

7. Sir William Watson Cheyne, *Lister and His Achievement* (London, 1925), p. 76.

8. Louis Pasteur, "Mémoire sur les Corpuscules Organisés qui Existent dans L'Atmosphère," *Annales de Chimie et de Physique*, 64 (1862), 28-29.

9. John Scott Burdon-Sanderson, "Remarks on the Attributes of the Germinal Particles of *Bacteria*, in reply to Prof. Tyndall," *Nature*, 17 (1877), 84.

10. Roberts, *op. cit.*, p. 458.

11. *Ibid.*, pp. 471-472.

12. Bulloch, *op. cit.*, p. 193.

13. Fielding Hudson Garrison, *An Introduction to the History of Medicine*, 4th Ed. (Philadelphia, 1929), p. 577.

14. Kenneth Vivian Thimann, *The Life of Bacteria*, 2nd Ed. (New York, 1963), p. 18.

15. Joseph Lister, "On the Germ Theory of Putrefaction and Other Fermentative Changes," *The Collected Papers of Joseph, Baron Lister*, (Oxford, 1909), vol. I, 288.

16. Bulloch, *op. cit.*, p. 195.

17. George Sykes, *Disinfection and Sterilization*, 2nd Ed. (London, 1965), p. 6.

Chapter 3

1. Charles Egolf Clifton, *Introduction to the Bacteria*, 2nd Ed., (New York, 1958), p. 391.

2. Henry Charlton Bastian, *Modes of Origin of the Lowest Organisms* (London, 1871), p. 4.

3. Henry Charlton Bastian wrote the following books on various aspects of the spontaneous generation controversy relating to the microbial world:

Modes of Origin of the Lowest Organisms (London, 1871).

The Beginnings of Life: Being Some Account of the Nature, Modes of Origin and Transformations of Lower Organisms, 2 vols. (London, 1872).

The Nature and Origin of Living Matter (London, 1905).

The Evolution of Life (London, 1907).

The Origin of Life: Being an Account of Experiments with Certain Superheated Saline Solutions in Hermetically Sealed Vessels (London, 1911).

4. Edwin Ray Lankester, "Dr. Bastian and Prof. Tyndall on Spontaneous Generation," *Nature*, 13 (1876), 324.

5. Henry Charlton Bastian, "Dr. Sanderson's Experiments and Archebiosis," *Nature*, 8 (1873), 200.

6. "Nearly twenty years ago a curious red fungus or mould *(Oïdium aurantiacum)* attacked the bread of Paris. M. Payen exposed pieces of bread, upon which spores of the fungus had been sown, for half an hour to $248°$ F. in tubes; the red fungus afterwards germinated, although its vitality was destroyed when the temperature was raised to $284°$ F." Quoted from: Edward Frankland, "On the Water Supply of the Metropolis," (Friday, March 29, 1867). *Proceedings of the Royal Institution of Great Britain*, 5 (1867), 126.

Also see: Anselme Payen, "Températures que peuvent supporter les sporules de l'*Oïdium aurantiacum* sans perdre leur faculté végétative," *Annales de Chimie et de Physique*, 24 (1848), 253-255. This is the paper to which Frankland referred in 1867.

7. Henry Charlton Bastian, "Reply to Professor Huxley's Inaugural Address at Liverpool on the Question of the Origin of Life," *Nature*, 2 (1870), 412.

8. William Roberts, "Studies on Biogenesis," *Philosophical Transactions of the Royal Society of London*, 164 (1874), 458 and 472.

9. Edward Frankland, "On the Proposed Water Supply for the Metropolis." (Friday, April 3, 1868). *Proceedings of the Royal Institution of Great Britain*, 5 (1868), 346-370.

10. Thomas Henry Huxley, "On the Relations of Penicillium, Torula, and Bacterium. (Special Report of an Address delivered in the Biological Section of the British Association for the Advancement of Science, Sept. 13th, 1870.)" *Quarterly Journal of Microscopical Science*, 10 (New Series) (1870), 359.

11. John Scott Burdon-Sanderson, "Dr. Bastian's Experiments on the Beginnings of Life," *Nature*, 7 (1873), 180-181.

12. For biographical information on John Burdon-Sanderson see: Ghetal Herschell Burdon-Sanderson, *Sir John Burdon Sanderson* (Oxford, 1911).

13. Burdon-Sanderson, *op. cit.*, p. 180.

14. Edwin Ray Lankester, "Dr. Sanderson's Experiments," *Nature*, 7 (1873), 242-243. See Lankester's request of Sanderson for information concerning the nature of cheese employed by Bastian.

15. Burdon-Sanderson, *op. cit.*, p. 180.

16. In the following chapter I shall discuss in detail the clash between Tyndall and Burdon-Sanderson over the latter's 1872 observation that neutralized turnip infusion (with *no* cheese) after boiling and sealing still contained: "Bacteria, about 0.003 mm. in length, which exhibited oscillatory movements." See *Nature*, 7 (1873), 181.

17. Burdon-Sanderson, *op. cit.*, p. 181.
18. *Ibid.*
19. Lankester, "Dr. Sanderson's Experiments," 243.
20. William Roberts, "Dr. Bastian's Experiments on the Beginning of Life," *Nature*, 7 (1873), 302.
21. Roberts, "Studies on Biogenesis," 462.
22. John Scott Burdon-Sanderson, "Dr. Bastian's Turnip-Cheese Experiments," *Nature*, 8, (1873), 141.
23. *Ibid.*
24. *Ibid.*, p. 142.
25. *Ibid.*
26. *Ibid.*, p. 143.
27. *Ibid.*
28. John Scott Burdon-Sanderson, "Note on Huizinga's Experiments on Abiogenesis," *Nature*, 8 (1873), 478.
29. Edwin Ray Lankester, "Experiments on the Development of Bacteria in Organic Infusions," *Nature*, 8 (1873), 504-505.
30. C.C. Pode and Edwin Ray Lankester, "Experiments on the Development of *Bacteria* in Organic Infusions," *Proceedings of the Royal Society of London*, 21 (1873), 358.
31. Edwin Ray Lankester, "An Experiment on the Destructive Effect of Heat upon the Life of Bacteria and their Germs," *Nature*, 9 (1874), 421.
32. For biographical information on John Tyndall see: Arthur Stewart Eve and C.H. Creasey, *Life and Work of John Tyndall* (London, 1945).
33. John Tyndall, "The Optical Deportment of the Atmosphere in Relation to the Phenomena of Putrefaction and Infection," *Philosophical Transactions of the Royal Society of London*, 166 (1876), 57.
34. Article written by "Inquirer" and published in the 10 February 1876 issue of *Nature*, 13 (1876), 285.
35. For further information on Ferdinand Julius Cohn see: Pauline Cohn, *Ferdinand Cohn* (Breslau, 1901); Hubert A. Lechevalier and Morris Solotorovsky, *Three Centuries of*

Microbiology (New York, 1965), pp. 68-69; G. Wunschmann, "Ferdinand Julius Cohn," *Allgemeine Deutsche Biographie*, 47 (1903), 503-505.

36. Ferdinand Julius Cohn, "Untersuchungen über Bacterien I," *Beiträge zur Biologie der Pflanzen*, Bd. 1, Heft 2 (1872), 145. [Publication date for Heft 2 is given in *Heinsius Allgemeines Bücher-Lexikon*, Bd. 15 (1877), 316. Note that the title page for Band 1 of the *Beiträge* bears the date 1875.]

Cohn's journal, *Beiträge zur Biologie der Pflanzen*, first appeared in 1870 when the first portion was published. A second part came out in 1872, and in 1875 a third segment appeared which completed the first volume of the journal. Volume II was published in 1877, and Volume III, also containing papers written during the 1870's appeared in 1883. Although all the volumes consist of three parts each, only in Volume I are page numbers separate for each part of the volume.

Reviews of each of the three parts of Volume I of Cohn's journal appeared in *Nature:* 26 January 1871 for Part I; 20 February 1873 for Part II; and 17 August 1876 for Part III.

37. *Ibid.*, p. 218.

38. Louis Pasteur, "Animalcules infusoires vivant sans gaz oxygène libre et déterminant des fermentations." *Comptes Rendus Hebdomadaires des Séances de l'Académie des Sciences*, 52 (1861) 344-347.

39. Cohn, *op. cit.*, p. 176.

40. Ferdinand Julius Cohn, "Untersuchungen über Bacterien II," *Beiträge zur Biologie der Pflanzen*, Bd. 1, Heft 3 (1875), 189.

41. *Ibid.*, p. 190.

42. *Ibid.*, pp. 194-195.

43. "Scientific Serials," *Nature*, 14 (1876), 202.

44. Ferdinand Julius Cohn, *Die Pflanze* (Breslau, 1897), vol. II, p. 472.

Chapter 4

1. John Scott Burdon-Sanderson, "Dr. Bastian's Experiments on the Beginnings of Life," *Nature*, 7 (1873), 180-181.
2. *Ibid.*, p. 180.
3. Ferdinand Julius Cohn, "Untersuchungen über Bacterien IV; Beiträge zur Biologie der Bacillen," *Beiträge zur Biologie der Pflanzen*, Bd. 2, Heft 2 (1876), 249-276. [Publication date for Heft 2 is given in *Christian Gottlob Kayser's Bücher-Lexikon*, Bd. 19 (1877), 96. Note that the title page for Band 2 of the *Beiträge* bears the date 1877.]
4. William Roberts, "Studies on Biogenesis," *Philosophical Transactions of the Royal Society of London*, 164 (1874), 457-477.
5. Ferdinand Julius Cohn, "Untersuchungen über Bacterien I," *Beiträge zur Biologie der Pflanzen*, Bd. 1, Heft 2 (1872), 127-224. [Publication date for Heft 2 is given in *Heinsius Allgemeines Bücher-Lexikon*, Bd. 15 (1877), 316. Note that the title page for Band 1 of the *Beiträge* bears the date 1875.]
6. Roberts, *op. cit.*, p. 473.
7. John Tyndall, "The Optical Condition of the Atmosphere in its Bearings on Putrefaction and Infection," *Proceedings of the Royal Institution of Great Britain*, 8 (1876), 6-27. John Tyndall's conclusions of early 1876 are also in: (a) "On the Optical Deportment of the Atmosphere in Reference to the Phenomena of Putrefaction and Infection," *Proceedings of the Royal Society of London*, 24 (1876) 171-183, and (b) "Prof. Tyndall on Germs," *Nature*, 13 (1876), 252-254 and 268-270.
8. Cohn, "Untersuchungen über Bacterien IV; Beiträge zur Biologie der Bacillen," (1876), 249-276.
9. John Tyndall, "The Optical Condition of the Atmosphere in its Bearings on Putrefaction and Infection," *Proceedings of the Royal Institution of Great Britain*, 8 (1876), 15-16.

10. *Ibid.*, p. 17.
11. *Ibid.*, p. 20.
12. Henry Charlton Bastian, "Prof. Tyndall on Germs," *Nature*, 13 (1876), 284-285. Also see two letters signed by "Inquirer" which may have been products of Bastian's pen: (a) *Nature*, 13 (1876), 285-286; and (b) *Nature*, 13 (1876), 347.
13. John Tyndall, "Prof. Tyndall on Germs," *Nature*, 13 (1876), 305. Also note Tyndall's use of a letter written by Louis Pasteur in order to support Tyndall's stand against Bastian — *Nature*, 13 (1876), 305-306.
14. John Scott Burdon-Sanderson, "Remarks on the Attributes of the Germinal Particles of *Bacteria*, in Reply to Prof. Tyndall," *Nature*, 17 (1877), 84-87.
15. John Tyndall, "The Optical Deportment of the Atmosphere in Relation to the Phenomena of Putrefaction and Infection," *Philosophical Transactions of the Royal Society of London*, 166 (1876), 71-72.
16. *Ibid.*, p. 57.
17. Cohn, "Untersuchungen über Bacterien IV; Beiträge zur Biologie der Bacillen," (1876), 254-257.
18. Ferdinand Julius Cohn, "Untersuchungen über Bacterien II," *Beiträge zur Biologie der Pflanzen*, Bd. 1, Heft 3 (1875), 189.
19. *Ibid.*, p. 196. Also see: Cohn, "Untersuchungen über Bacterien IV; Beiträge zur Biologie der Bacillen," (1876), 267-268.
20. Cohn, "Untersuchungen über Bacterien IV; Beiträge zur Biologie der Bacillen," (1876), 267-268.
21. Burdon-Sanderson, "Dr. Bastian's Experiments on the Beginnings of Life," 180.
22. Cohn, "Untersuchungen über Bacterien IV; Beiträge zur Biologie der Bacillen," (1876), 259.
23. *Ibid.*
24. John Tyndall, "Further Researches on the Deportment and Vital Persistence of Putrefactive and Infective

Organisms from a Physical Point of View," *Philosophical Transactions of the Royal Society of London*, 167 (1877), 149.

25. Cohn, "Untersuchungen über Bacterien IV; Beiträge zur Biologie der Bacillen," (1876), 267-268.

26. For an excellent account of the meeting between Cohn and Koch in spring, 1876, and of Cohn's generosity, read: Hubert A. Lechevalier and Morris Solotorovsky, *Three Centuries of Microbiology* (New York, 1965), pp. 68-69.

27. For biographical information on Robert Koch see: Richard Max Emil Julius Theodor Bochalli, *Robert Koch, Der Schöpfer Der Modernen Bakteriologie* (Stuttgart, 1954). Rudolf Harms, *Robert Koch: Arzt und Forscher* (Hamburg, 1966). Émile Lagrange, *Robert Koch: Sa Vie et Son Oeuvre* (Tours, 1938). Hubert A. Lechevalier and Morris Solotorovsky, *Three Centuries of Microbiology* (New York, 1965). Elie Metchnikoff, *The Founders of Modern Medicine* (New York, 1939). Bernhard Möllers, *Robert Koch: Persönlichkeit und Lebenswerk* (Hanover, 1950).

28. Cohn, "Untersuchungen über Bacterien IV; Beiträge zur Biologie der Bacillen," (1876), 275.

29. Pauline Cohn, *Ferdinand Cohn* (Breslau, 1901), pp. 203-204.

30. John Tyndall, "Further Researches on the Deportment and Vital Persistence of Putrefactive and Infective Organisms from a Physical Point of View," *Philosophical Transactions of the Royal Society of London*, 167 (1877), 152.

31. John Tyndall, "Fermentation and Its Bearings on Surgery and Medicine," *Fragments of Science*, (New York, 1898), vol. 2, pp. 251 and 281.

32. John Tyndall, "Further Researches on the Deportment and Vital Persistence of Putrefactive and Infective Organisms from a Physical Point of View," *Philosophical Transactions of the Royal Society of London*, 167 (1877), 159.

33. *Ibid.*, pp. 172-173.

34. *Ibid.*, pp. 174-175.

35. John Tyndall, "On Heat as a Germicide when Discontinuously Applied," [The article consists of Tyndall's letter to Huxley.] *Proceedings of the Royal Society of London*, 25 (1877), 569-570.

36. John Tyndall, "Further Researches on the Deportment and Vital Persistence of Putrefactive and Infective Organisms from a Physical Point of View," *Philosophical Transactions of the Royal Society of London*, 167 (1877), 149.

37. *Ibid.*, p. 151.

38. John Scott Burdon-Sanderson, "Remarks on the Attributes of the Germinal Particles of *Bacteria*, in Reply to Prof. Tyndall," *Nature*, 17 (1877), 86-87.

39. John Tyndall, "Spontaneous Generation," *Fragments of Science*, 2, (New York, 1898), pp. 290-334. See reference to Cohn on p. 300 — no mention of bacterial endospore formation. See four places where Tyndall might have mentioned Cohn: pp. 318, 323, 329, and 332-333.

40. John Tyndall, *New Fragments* (New York, 1898), pp. 190-192. See Tyndall's 1884 Introduction to the 1885 English translation by Tyndall's wife, Lady Claud Hamilton, of René Vallery-Radot's book: *Louis Pasteur: His Life and Labours* (New York, 1885).

41. William T. Jeans, *Lives of the Electricians* (London, 1887), pp. 98-99.

Chapter 5

1. Henry Charlton Bastian, "The Fermentation of Urine and the Germ Theory," *Nature*, 14 (1876), 309-311.

2. John Tyndall, "Preliminary Note on the Deportment of Organisms in Organic Infusions," *Proceedings of the Royal Society of London*, 25 (1877), 503-506.

3. For biographical information on Louis Pasteur see:
René Jules Dubos, *Louis Pasteur, Free Lance of Science* (Boston, 1950).
René Jules Dubos, *Pasteur and Modern Science* (Garden City, N.Y., 1960).
Émile Duclaux, *Pasteur, the History of a Mind*, Translated by Erwin F. Smith and Florence Hedges, (Philadelphia, 1920).
René Vallery-Radot, *The Life of Pasteur*, Translated by Mrs. R.L. Devonshire, (New York, 1960 — Dover edition of the 1901 edition).

4. Louis Pasteur, "Mémoire sur les Corpuscules Organisés qui Existent dans l'Atmosphère," *Annales de Chimie et de Physique*, 64 (1862), 62.

5. William Roberts, "Note on the Influence of Liquor Potassae and an Elevated Temperature on the Origin and Growth of Microphytes," *Nature*, 15 (1877), 302.

6. John Tyndall, "Note on the Deportment of Alkalized Urine," *Nature*, 15 (1877), 303.

7. Louis Pasteur, "The Spontaneous Generation Question," *Nature*, 15 (1877), 380-381.

8. Henry Charlton Bastian, "The Spontaneous Generation Question," *Nature*, 15 (1877), 381.

9. Pasteur, "The Spontaneous Generation Question," 381.

10. Henry Charlton Bastian, "The Commission of the French Academy and the Pasteur-Bastian Experiments," *Nature*, 16 (1877), 277.

11. *Ibid.*, p. 278.

12. The third member of the Commission was originally Jean Baptiste Joseph Dieudonné Boussingault (1802-1887). However, Boussingault was unable to serve; his post was filled by Van Tieghem.

13. Bastian, "The Commission of the French Academy and the Pasteur-Bastian Experiments," 279.

14. John Tyndall, *New Fragments*, (New York, 1898), pp. 190-192. Read Tyndall's 1884 Introduction to the 1885

English translation by Tyndall's wife, Lady Claud Hamilton, of René Vallery-Radot's book: *Louis Pasteur: His Life and Labours*, (New York, 1885).

15. Charles Edouard Chamberland, "Résistance des germes de certains organismes à la température de 100 degrés; conditions de leur développement," *Comptes Rendus de l'Académie des Sciences*, 88 (1879), 659-661.

16. William Bulloch, *The History of Bacteriology*, (London, 1938), p. 109.

Chapter 6

1. John Tyndall, "Prof. Tyndall on Germs," *Nature*, 13 (1876), 305.

2. Henry Charlton Bastian, "Prof. Tyndall on Germs," *Nature*, 13 (1876), 285.

3. John Tyndall, "Further Researches on the Deportment and Vital Persistence of Putrefactive and Infective Organisms from a Physical Point of View," *Philosophical Transactions of the Royal Society of London*, 167 (1877), 204-205.

4. William Roberts, "Studies on Biogenesis," *Philosophical Transactions of the Royal Society of London*, 164 (1874), 464-465.

5. Martin Frobisher, *Fundamentals of Microbiology*, 7th Ed., (Philadelphia, 1962), p. 409.

6. George Sykes, *Disinfection and Sterilization*, 2nd Ed., (London, 1965), pp. 110-111.

7. Frobisher, *op. cit.*, pp. 152-153.

8. *Ibid.*, p. 409.

9. Roberts, *op. cit.*, p. 464.

10. Tyndall, "Further Researches," 229-230.

11. The degree of acidity or alkalinity of any solution is generally expressed by a value from a pH scale ranging from 1 (very acid) through 7 (neutral) up to 14 (very alkaline). The pH reactions of some common substances are: vinegar 2.3; fresh milk 6.8; human blood 7.4; and milk of magnesia 11.0.

12. "Societies and Academies," *Nature*, 14 (1876), 284.
13. Tyndall, "Further Researches," 188-189.
14. Roberts, *op. cit.*, p. 465.
15. Frobisher, *op. cit.*, p. 281.
16. Kenneth Vivian Thimann, *The Life of Bacteria*, 2nd Ed., (New York, 1963), p. 183.
17. *Ibid.*, p. 167.
18. Sykes, *op. cit.*, p. 118.
19. Frobisher, *op. cit.*, p. 156.
20. Roberts, *op. cit.*, pp. 465 and 473.
21. Frobisher, *op. cit.*, p. 409.
22. Thimann, *op. cit.*, pp. 43, 44, 132. Also see: Edward Emanuel Klein, *Micro-Organisms and Disease*, 3rd Ed., (London, 1886), pp. 114-115.
23. *Ibid.*, Thimann, pp. 43, 44, 132; Klein, pp. 108-110.
24. Charles Egolf Clifton, *Introduction to the Bacteria*, 2nd Ed., (New York, 1958), p. 69.
25. John Tyndall, *Essays on the Floating-Matter of the Air in Relation to Putrefaction and Infection*, (New York, 1888), p. 107.
26. Roger Y. Stanier, Michael Doudoroff, and Edward A. Adelberg, *The Microbial World*, 2nd Ed., (Englewood Cliffs, N.J., 1963), p. 356.
27. Sykes, *op. cit.*, p. 109.
28. Stanier *et al.*, *op. cit.*, p. 355.
29. Thimann, *op. cit.*, pp. 183 and 185.
30. Stanier *et al.*, *op. cit.*, p. 356.
31. Frobisher, *op. cit.*, p. 153.

Chapter 7

1. George Miller Sternberg, *Bacteria*, (New York, 1884), p. 168.
2. William Burrows, *Textbook of Microbiology*, 18th Ed., (Philadelphia, 1963), p. 208.
3. George Sykes, *Disinfection and Sterilization*, 2nd Ed., (London, 1965), p. 109.

4. Martin Frobisher, *Fundamentals of Microbiology*, 7th Ed., (Philadelphia, 1962), p. 280. Also see: Sykes, *Ibid.*, *p. 110*.

5. John Tyndall, "Preliminary Note on the Development of Organisms in Organic Infusions," *Proceedings of the Royal Society of London*, 25 (1877), 503-506.

6. William Roberts, "Studies on Biogenesis," *Philosophical Transactions of the Royal Society of London*, 164 (1874), 460.

7. *Ibid.*, pp. 461 and 473.

8. Joseph Lister, "A Further Contribution to the Natural History of Bacteria and the Germ Theory of Fermentative Changes," [1873] *The Collected Papers of Joseph, Baron Lister*, (Oxford, 1909), Vol. 1, pp. 311-312.

9. Sykes, *op. cit.*, p. 109.

10. An apparatus devised in 1681 by Denys Papin (1647-1712). Abraham Wolf, *A History of Science, Technology and Philosophy in the 16th and 17th Centuries*, 2nd Ed., (New York, 1959), Vol. II, pp. 548-549.

11. John Scott Burdon-Sanderson, "Note on Huizinga's Experiments on Abiogenesis," *Nature*, 8 (1873), 478.

12. Roberts, *op. cit.*, pp. 461 and 465.

13. Sternberg, *op. cit.*, p. 169.

14. Charles Sumner Dolley, *The Technology of Bacteria Investigation*, (Boston, 1885), p. 61.

15. William Bulloch, *The History of Bacteriology*, (London, 1938), p. 234.

16. Frobisher, *op. cit.*, p. 279.

17. Burrows, *op. cit.*, p. 61.

18. Sykes, *op. cit.*, p. 115.

19. John Tyndall, "On Heat as a Germicide when Discontinuously Applied," *Proceedings of the Royal Society of London*, 25 (1877), 569-570.

20. See for example: Roger Y. Stanier, Michael Doudoroff, and Edward A. Adelberg, *The Microbial World*, 2nd Ed., (Englewood Cliffs, N.J., 1963), p. 17.

21. John Tyndall, "Further Researches on the Deportment and Vital Persistence of Putrefactive and Infective Organisms from a Physical Point of View," *Philosophical Transactions of the Royal Society of London*, 167 (1877), 194-195.

22. Stanier *et al.*, *op. cit.*, p. 17.

23. Frobisher, *op. cit.*, p. 279.

24. Sykes, *op. cit.*, p. 141.

25. John Scott Burdon-Sanderson, "Dr. Bastian's Experiments on the Beginnings of Life," *Nature*, 7 (1873), 181.

26. Roberts, *op. cit.*, p. 467.

27. Sykes, *op. cit.*, p. 142.

28. Joseph Lister, "On the Lactic Fermentation and Its Bearings on Pathology," [1878] *The Collected Papers of Joseph, Baron Lister*, (Oxford, 1909), Vol. 1, pp. 354-356.

29. Alexander Crever Abbott, *The Principles of Bacteriology*, 6th Ed., (Philadelphia, 1902), p. 63.

30. Frobisher, *op. cit.*, p. 140.

31. Sykes, *op. cit.*, p. 110.

32. Stanier *et al.*, *op. cit.*, p. 18.

33. Frobisher, *op. cit.*, p. 280.

Bibliography

1. Abbott, Alexander Crever. *The Principles of Bacteriology.* 6th Ed. Philadelphia, 1902.

2. Ackerknecht, Erwin Heinz. *A Short History of Medicine.* New York, 1955.

3. Bary, Anton de. *Vergleichende Morphologie und Biologie der Pilze, Mycetozoen, und Bacterien.* Leipzig, 1884.

4. ————. *Comparative Morphology and Biology of the Fungi, Mycetozoa, and Bacteria.* Translated by Henry E.F. Garnsey. Revised by Isaac Bayley Balfour. Oxford, 1887.

5. ————. *Vorlesungen über Bacterien.* 2nd Ed. Leipzig, 1887.

6. ————. *Lectures on Bacteria.* 2nd Ed. Translated by Henry E.F. Garnsey. Revised by Isaac Bayley Balfour. Oxford, 1887.

7. Bastian, Henry Charlton. "Facts and Reasonings Concerning the Heterogeneous Evolution of Living Things," *Nature*, 2 (1870), 170-177; 193-201; 219-228.

8. ————. "Reply to Prof. Huxley's Inaugural Address at Liverpool on the Question of the Origin of Life," *Nature*, 2 (1870), 410-413; 431-434.

9. ———. *The Modes of Origin of Lowest Organisms: Including A Discussion of the Experiments of M. Pasteur, And a Reply To Some Statements by Professors Huxley and Tyndall.* London, 1871.

10. ———. "On Some Heterogenetic Modes of Origin of Flagellated Monads, Fungus-germs, and Ciliated Infusoria," *Proceedings of the Royal Society of London*, 20 (1872), 239-264.

11. ———. *The Beginnings of Life: Being Some Account of the Nature, Modes of Origin and Transformations of Lower Organisms.* 2 vols. London, 1872.

12. ———. "Dr. Sanderson's Experiments and Archebiosis," *Nature*, 8 (1873), 199-200.

13. ———. "Note on the Origin of *Bacteria*, and on Their Relation to the Process of Putrefaction," [1872.] *Proceedings of the Royal Society of London*, 21 (1873), 129-131.

14. ———. "On the Temperature at which *Bacteria*, *Vibriones*, and Their Supposed Germs are Killed When Immersed in Fluids or Exposed to Heat in a Moist State," *Proceedings of the Royal Society of London*, 21 (1873), 224-232.

15. ———. "Further Observations on the Temperature at Which *Bacteria*, *Vibriones*, and Their Supposed Germs are Killed When Exposed to Heat in a Moist State; and on the Causes of Putrefaction and Fermentation," *Proceedings of the Royal Society of London*, 21 (1873), 325-338.

16. ———. "Spontaneous Generation," *Nature*, 9 (1873), 482-483.

17. ———. *Evolution and the Origin of Life*. London, 1874.

18. ———. "The Microscopic Germ Theory of Disease," *The Monthly Microscopical Journal: Transaction of the Royal Microscopical Society, and Record of Histological Research*, 14 (1875), 65-79; 129-140.

19. ———. "The Germ Theory of Disease," *Transactions of the Pathological Society of London*, 26 (1875), 255-284; 334-345.

20. ———. "Prof. Tyndall on Germs," *Nature*, 13 (1876), 284-285.

21. ———. "The Fermentation of Urine and the Germ Theory," *Nature*, 14 (1876), 309-311.

22. ———. "Influence des Forces Physico-Chimiques sur les Phénomènes de Fermentation," *Comptes Rendus hebdomadaires des Séances de l'Académie des Sciences*, 83 (1876), 159-161.

23. ———. "Note sur la Fermentation de l'Urine, à propos d'une Communication de M. Pasteur," *Comptes Rendus hebdomadaires des Séances de l'Académie des Sciences*, 83 (1876), 362-363.

24. ———. "Sur la Fermentation de l'Urine, Réponse à M. Pasteur," *Comptes Rendus hebdomadaires des Séances de l'Académie des Sciences*, 83 (1876), 488-490. Also see: 84 (1877), 187-190; 306-367.

25. ———. "Researches Illustrative of the Physico-Chemical Theory of Fermentation, and of the Conditions Favouring Archebiosis in Previously Boiled Fluids," *Proceedings of the Royal Society of London*, 25 (1876), 149-156.

26. ———. "The Spontaneous Generation Question," *Nature*, 15 (1877), 380-381.

27. ———. "The Commission of the French Academy and the Pasteur-Bastian Experiments," *Nature*, 16 (1877), 276-279.

28. ———. "On the Conditions Favouring Fermentation and the Appearance of Bacilli, Micrococci, and Torulae in Previously Boiled Fluids," *The Journal of the Linnean Society* (Zoology), 14 (1877), 89-93.

29. ———. *Studies in Heterogenesis*. London, 1903. [1901-03].

30. ———. *The Nature and the Origin of Living Matter*. London, 1905.

31. ———. *The Evolution of Life*. London, 1907.

32. ———. *The Origin of Life: Being an Account of Experiments with Certain Superheated Saline Solutions in Hermetically Sealed Vessels*. London, 1911.

33. Bochalli, Richard Max Emil Julius Theodor. *Robert Koch (Der Schöpfer Der Modernen Bakteriologie)*. Stuttgart, 1954.

34. Bulloch, William. *The History of Bacteriology*. London, 1938.

35. Burdon-Sanderson, Ghetal Herschell. *Sir John Burdon-Sanderson*. Oxford, 1911.

36. Burdon-Sanderson, John Scott. "The Origin and Distribution of Microzymes (Bacteria) in Water, and the Circumstances Which Determine Their Existence in the Tissues and

Liquids of the Living Body," *Quarterly Journal of Microscopical Science*, 11 (1871), 323-352.

37. ―――. "Connection Between Pyaemia and Bacteria," *Monthly Microscopical Journal*, 8 (1872), 35-36.

38. ―――. "Dr. Bastian's Experiments on the Beginnings of Life," *Nature*, 7 (1873), 180-181.

39. ―――. "Dr. Bastian's Turnip-Cheese Experiments," *Nature*, 8 (1873), 141-143.

40. ―――. "Note on Huizinga's Experiments on Abiogenesis," *Nature*, 8 (1873), 478-479. Also see: *Report of the British Association for the Advancement of Science*, 43 (1873), *(Sect.)*, 131-133.

41. ―――. [Remarks in the Discussion on the Germ Theory of Disease.] *Transactions of the Pathological Society of London*, 26 (1875), 284-289.

42. ―――. "Remarks on the Attributes of the Germinal Particles of *Bacteria*, in Reply to Prof. Tyndall," *Proceedings of the Royal Society of London*, 26 (1877), 416-426. Also see: *Nature*, 17 (1877), 84-87.

43. Burrows, William. *Textbook of Microbiology*. 18th Ed. Philadelphia, 1963.

44. Cameron, Sir Hector Clare. *Reminiscences of Lister*. Glasgow, 1927.

45. Chamberland, Charles Edouard. "Résistance des Germes de Certains Organismes à la Température de 100 Degrés; Conditions de leur Développement," *Comptes Rendus hebdomadaires des Séances de l'Académie des Sciences*, 88 (1879), 659-661.

46. Cheyne, Sir William Watson. *Lister and His Achievement*. London, 1925.

47. Clark, Paul Franklin. *Pioneer Microbiologists of America*. Madison, 1961.

48. Clifton, Charles Egolf. *Introduction to the Bacteria*. 2nd Ed. New York, 1958.

49. Cohn, Ferdinand Julius. *Ueber Bacterien, Die Kleinsten Lebenden Wesen*. Berlin, 1872. Cohn's monograph may be found in: *Sammlung Gemeinverständlicher Wissenschftlicher Vorträge*. 7th Series containing Numbers 145-168 (1872-73), Number 165, 3-35.

50. ———. *Bacteria: The Smallest of Living Organisms*, (1872). Translated by Charles Sumner Dolley (1881). Baltimore, 1939.

51. ———. "Untersuchungen über Bacterien I," *Beiträge zur Biologie der Pflanzen*, Bd. 1, Heft 2 (1872), 127-224. [Publication date for Heft 2 is given in *Heinsius Allgemeines Bücher-Lexikon*, Bd. 15 (1877), 316. Note that the title page for Band 1 of the *Beiträge* bears the date 1875.]

52. ———. "Ueber Bacterien und deren Beziehungen zur Faulniss und zu Contagien," *Quarterly Journal of Microscopical Science*, 12 (1872), 207-210.

53. ———. "Researches on Bacteria," (Translated) *Quarterly Journal of Microscopical Science*, 13 (1873), 156-163.

54. ———. "Untersuchungen über Bacterien II," *Beiträge zur Biologie der Pflanzen*, Bd. 1, Heft 3 (1875), 141-207.

55. ———. "Untersuchungen über Bacterien IV; Beiträge zur Biologie der Bacillen," *Beiträge zur Biologie der Pflanzen*,

Bd. 2, Heft 2 (1876), 249-276. [Publication date for Heft 2 is given in *Christian Gottlob Kayser's Bücher-Lexikon*, Bd. 19 (1877), 96. Note that the title page for Band 2 of the *Beiträge* bears the date 1877.]

56. ———. *Die Pflanze.* 2 vols. Breslau, 1901.

57. Cohn, Pauline. *Ferdinand Cohn.* Breslau, 1901.

58. Conant, James Bryant, ed. *Pasteur's Study of Fermentation.* (Case no. 6 of Harvard Case Histories in Experimental Science). Cambridge, Mass., 1952.

59. ———. *Pasteur's and Tyndall's Study of Spontaneous Generation.* (Case no. 7 of Harvard Case Histories in Experimental Science). Cambridge, Mass., 1959.

60. Crace-Calvert, Frederick, "On the Development of Germ-Life," *Report of the British Association for Advancement of Science*, 40 (1870) *(Sect.)*, 132-133.

61. ———. "On Protoplasmic Life," *Proceedings of the Royal Society of London*, 19 (1871), 468-472.

62. ———. "On Putrefaction," *Proceedings of the Royal Society of London*, 20 (1872), 185-187.

64. ———. "On the Relative Power of Various Substances in Preventing Putrefaction and the Development of Protoplasmic and Fungus-Life," *Proceedings of the Royal Society of London*, 20 (1872), 187-191.

65. ———. "On the Relative Power of Various Substances in Arresting Putrefaction and the Development of Protoplasmic and Fungus-Life," *Proceedings of the Royal Society of London*, 20 (1872), 191-192.

66. Dallinger, William Henry. "Professor Tyndall's Experiments on Spontaneous Generation and Dr. Bastian's Position," *Popular Science Review*, 15 (1876), 113-127.

67. ———. "On the Life-History of a Minute Septic Organism: With an Account of Experiments Made to Determine its Thermal Death Point," *Proceedings of the Royal Society of London*, 27 (1878), 332-350.

68. Dawes, Benjamin. *A Hundred Years of Biology*. London, 1952.

69. De Kruif, Paul Henry. *Microbe Hunters*. New York, 1926.

70. Doetsch, Raymond Nicholas, ed. *Microbiology (Historical Contributions from 1776 to 1908)*. New Brunswick, N.J., 1960.

71. Dolley, Charles Sumner. *The Technology of Bacteria Investigation*. Boston, 1885.

72. Downes, Arthur and Thomas P. Blunt. "Researches on the Effect of Light upon *Bacteria* and other Organisms," *Proceedings of the Royal Society of London*, 26 (1877), 488-500.

73. ———. "The Influence of Light upon Bioplasm," *Nature*, 18 (1878), 398-399.

74. ———. "On the Influence of Light upon Protoplasm," *Proceedings of the Royal Society of London*, 28 (1878), 199-212.

75. Drysdale, John James, "On the Germ Theories of Infectious Diseases," *Proceedings of the Literary and Philosophical Society of Liverpool*, 33 (1878), 1-74.

76. Dubos, René Jules. *Louis Pasteur, Free Lance of Science.* Boston, 1950.

77. ———. *Pasteur and Modern Science.* Garden City, New York, 1960.

78. Duclaux, Émile. *Pasteur, Histoire d'un Esprit.* Sceaux, 1896.

79. ———. *Pasteur, The History of a Mind.* Translated by Erwin F. Smith and Florence Hedges. Philadelphia, 1920.

80. Dukes, Cuthbert Esquire. *Lord Lister.* London, 1924.

81. Eve, Arthur Stewart and C.H. Creasey. *Life and Work of John Tyndall.* London, 1945.

82. Ewart, J. Cossar. "The Life-History of *Bacterium termo* and *Micrococcus*, with Further Observations on *Bacillus*," *Proceedings of the Royal Society of London*, 27 (1878), 474-480.

83. ———. "On the Life History of *Bacillus anthracis*," *Quarterly Journal of Microscopical Science*, 18 (1878), 161-170.

84. Farmer, Laurence. *Master Surgeon.* New York, 1962.

85. Fischer, Alfred. *The Structure and Functions of Bacteria.* Translated by A. Coppen Jones. Oxford, 1900.

86. Ford, William Webber. *Bacteriology.* New York, 1939.

87. Frankland, Edward, "On the Water Supply of the Metropolis," *Proceedings of the Royal Institution of Great Britain*, 5 (1867), 109-126.

88. ─────. "On the Proposed Water Supply for the Metropolis," *Proceedings of the Royal Institution of Great Britain*, 5 (1868), 346-370.

89. ─────. "Experiments on Spontaneous Generation," *Nature*, 3 (1871), 225.

90. Frobisher, Martin. *Fundamentals of Microbiology.* 7th Ed. Philadelphia, 1962.

91. Gage, Simon Henry. *The Microscope.* 17th Ed., Ithaca, N.Y., 1941.

92. Galdston, Iago. *Progress in Medicine.* New York, 1940.

93. Garrison, Fielding Hudson. *An Introduction to the History of Medicine.* 4th Ed. Philadelphia, 1929.

94. Godlee, Sir Rickman John. *Lord Lister.* 2nd Ed. London, 1918.

95. Gorton, David Allyn. *The History of Medicine.* 2 vols. New York, 1910.

96. Gradle, Henry. *Bacteria and the Germ Theory of Disease.* Chicago, 1883.

97. Gregory, Philip Herries. *The Microbiology of the Atmosphere.* New York, 1961.

98. Guthrie, Douglas. *Lord Lister.* Baltimore, 1949.

99. Haagensen, Cushman Davis and Wyndham E.G. Lloyd. *A Hundred Years of Medicine.* New York, 1943.

100. Harms, Rudolf. *Robert Koch: Arzt und Forscher.* Hamburg, 1966.

101. Hartley, Walter Noel. "Experiments Concerning the Evolution of Life from Lifeless Matter," [1871] *Proceedings of the Royal Society of London*, 20 (1872), 140-157.

102. Heymann, Bruno. *Robert Koch*. 2 vols. Leipzig, 1932.

103. Holmes, Oliver Wendell. *Medical Essays (1842-1882)*. Boston, 1891.

104. Hueppe, Ferdinand Adolph Theophil. *The Methods of Bacteriological Investigation*. Translated by Hermann M. Biggs. New York, 1886.

105. ———. *The Principles of Bacteriology*. Translated by E. O. Jordon. Chicago, 1899.

106. Huizinga, Dirk. "Experiments on Abiogenesis," *The Monthly Microscopical Journal: Transactions of the Royal Microscopical Society*, 10 (1873), 230-232.

107. ———. "New Experiments on Abiogenesis," *Nature*, 7 (1873), 380-381.

108. ———. "Additional Remarks on Abiogenesis," *Nature*, 8 (1873), 85-86.

109. Huxley, Thomas Henry. "Address to the British Association at Liverpool," *Report of the British Association for the Advancement of Science*, 40 (1870), lxxiii-lxxix.

110. ———. "On the Relations of Penicillium, Torula, and Bacterium. (Special Report of an Address delivered in the Biological Section of the British Association for the Advancement of Science, Sept. 13th, 1870." *Quarterly Journal of Microscopical Science*, 10 (New Series) (1870), 355-362.

111. ———. "The Border Territory Between the Animal and the Vegetable Kingdoms," *Proceedings of the Royal Institution of Great Britain*, 8 (1876), 28-34.

112. ———. [Address to the International Medical Congress, London, 1881.] "The Connection of the Biological Sciences with Medicine," *Nature*, 24 (1881), 342-346.

113. ———. *Discourses Biological and Geological. (Essays by Thomas Henry Huxley).* New York, 1896.

114. "Inquirer," *Nature*, 13 (1876), 285-287; 347.

115. Jeans, William T. *Lives of the Electricians: Professors Tyndall, Wheatstone, and Morse.* London, 1887.

116. Klein, Edward Emanuel. *Micro-Organisms and Disease.* 3rd Ed., London, 1886.

117. Koch, Robert. "Untersuchungen über Bacterien V; Die Aetiologie der Milzbrand-Krankheit, begrundet auf die Entwicklungsgeschichte des Bacillus anthracis," *Beiträge zur Biologie der Pflanzen*, Bd. 2, Heft 2 (1876), 277-310. [Publication date for Heft 2 is given in *Christian Gottlob Kayser's Bücher-Lexikon*, Bd. 19 (1877), 96. Note that the title page for Band 2 of the *Beiträge* bears the date 1877.]

118. ———. "Untersuchungen über Bacterien VI; Verfahren zur Untersuchung, zum Conserviren und Photographiren der Bacterien," *Beiträge zur Biologie der Pflanzen*, Bd. 2, Heft 3 (1877), 399-434.

119. ———. *Untersuchungen über die Aetiologie der Wundinfectionskrankheiten.* Leipzig, 1878.

120. ———. *Investigations into the Etiology of Traumatic*

Infective Diseases. Translated by William Watson Cheyne. London, 1880.

121. ———. *Gesammelte Werke von Robert Koch.* 3 vols. Leipzig, 1912.

122. Lagrange, Émile. *Robert Koch (Sa Vie et Son Oeuvre).* Tours, 1938.

123. Lambert, Royston, *Sir John Simon (1816-1904) and English Social Administration.* London, 1963.

124. Lankester, Edwin Ray. "Dr. Sanderson's Experiments," *Nature*, 7 (1873), 242-243.

125. ———. "Experiments on the Development of Bacteria in Organic Infusions," *Nature*, 8 (1873), 504-505.

126. ———. "An Experiment on the Destructive Effect of Heat upon the Life of Bacteria and their Germs," *Nature*, 9 (1874), 421-422.

127. ———. "Dr. Bastian and Prof. Tyndall on Spontaneous Generation," *Nature*, 13 (1876), 324.

128. Lechevalier, Hubert A. and Morris Solotorovsky. *Three Centuries of Microbiology.* New York, 1965.

129. Lister, Joseph, Baron. "On the Germ Theory of Putrefaction and Other Fermentative Changes," *Transactions of the Royal Society of Edinburgh*, 27 (1873), 313-344. Also see: *Nature*, 8 (1873), 212-214.

130. ———. "A Further Contribution to the Natural History of Bacteria and the Germ Theory of Fermentative Changes," *Quarterly Journal of Microscopical Science*, 13 (1873), 380-408.

131. ———. "On the Nature of Fermentation. The Introductory Address Delivered in King's College, London, October 1st, 1877," *Quarterly Journal of Microscopical Science*, 18 (1878), 177-194.

132. ———. "On the Lactic Fermentation, and Its Bearings on Pathology," *Transactions of the Pathological Society of London*, 29 (1878), 425-467.

133. ———. "On the Relation of Micro-organisms to Disease," *Quarterly Journal of Microscopical Science*, 21 (1881), 330-342.

134. ———. *The Collected Papers of Joseph, Baron Lister.* 2 vols. Oxford, 1909.

135. Locy, William Albert. *Biology and Its Makers.* New York, 1908.

136. Metchnikoff, Elie. *The Founders of Modern Medicine.* New York, 1939.

137. Migula, Walter. *Bacteriologisches Practicum.* Karlsruhe, 1892.

138. Miquel, Pierre. *Les Organismes Vivants de l'Atmosphère.* Paris, 1883.

139. Möllers, Bernhard. *Robert Koch: Persönlichkeit und Lebenswerk.* Hannover, 1950.

140. Newsholme, Sir Arthur. *Evolution of Preventive Medicine.* London, 1927.

141. ———. *The Story of Modern Preventive Medicine.* Baltimore, 1929.

142. Nicolle, Jacques. *Louis Pasteur, The Story of His Major Discoveries*, New York, 1961.

143. Nordenskiöld, Erik. *The History of Biology*. New York, 1928.

144. Osler, Sir William. *The Evolution of Modern Medicine*. New Haven, 1921.

145. Pasteur, Louis. "Animalcules infusoires vivant sans gaz ozygène libre et déterminant des fermentations." *Comptes Rendus hebdomadaires des Séances de l'Académie des Sciences*, 52 (1861), 344-347.

146. ———. "Mémoire sur les Corpuscules Organisés qui Existent dans l'Atmosphère," *Annales de Chimie et de Physique*, 64 (1862), 5-110.

147. ———. *Études sur le Vin*. 2nd Ed. Paris, 1873.

148. ———. *Études sur la Bière*. Paris, 1876.

149. ———. "Note sur l'altération de l'urine, à propos d'une communication du Dr. Bastian, de Londres," *Comptes Rendus hebdomadaires des Séances de l'Académie des Sciences*, 83 (1876), 176-180.

150. ———. "Sur l'altération de l'urine, Réponse à M. le Dr. Bastian," *Comptes Rendus hebdomadaires des Séances de l'Académie des Sciences*, 83 (1876), 377-378.

151. ———. "Note au sujet de l'expérience du Dr. Bastian, relative à l'urine neutralisée par la potasse," *Comptes Rendus hebdomadaires des Séances de l'Académie des Sciences*, 85 (1877), 178-180.

152. ―――. "The Spontaneous Generation Question," *Nature*, 15 (1877), 380-381.

153. ――― and Jules François Joubert. "Sur la fermentation de l'urine," *Comptes Rendus hebdomadaires des Séances de l'Académie des Sciences*, 83 (1876), 5-8, 10.

154. ――― ―――. "Note sur l'altération de l'urine, à propos des communications récentes du Dr. Bastian," *Comptes Rendus hebdomadaires des Séances de l'Académie des Sciences*, 84 (1877), 64-66, 206.

155. ――― ―――. "Sur les germes des bactéries en suspension dans l'atmosphère et dans les eaux," *Comptes Rendus hebdomadaires des Séances de l'Académie des Sciences*, 84 (1877), 206-209.

156. ――― and John Tyndall. *Les Microbes Organisés — Leur Rôle dans la Fermentation, la Putréfaction et la Contagion*. Paris, 1878.

157. Payen, Anselme. "Températures que peuvent supporter les sporules de l'*Oïdium aurantiacum* sans perdre leur faculté végétative," *Annales de Chimie et de Physique*, 24 (1848), 253-255.

158. Perkin, William Henry. "On the Newest Colouring Matters," *Proceedings of the Royal Institution of Great Britain*, 5 (1869), 566-574.

159. Pode, C. C. and Edwin Ray Lankester. "Experiments on the Development of *Bacteria* in Organic Infusions," *Proceedings of the Royal Society of London*, 21 (1873), 349-358.

160. Rayleigh, John William Strutt (Lord). "Discourse on The Scientific Work of Tyndall," *Proceedings of the Royal Institution of Great Britain*, 14 (1894), 216-224.

Bibliography

161. Roberts, William. "Dr. Bastian's Experiments on the Beginning of Life," *Nature*, 7 (1873), 302.

162. ―――. [Experiments on the Question of Biogenesis.] *Monthly Microscopical Journal*, 9 (1873), 228-230.

163. ―――. "Studies on Biogenesis," *Philosophical Transactions of the Royal Society of London*, 164 (1874), 457-477.

164. ―――. "Note on the Influence of Liquor Potassae and an Elevated Temperature on the Origin and Growth of Microphytes," *Proceedings of the Royal Society of London*, 25 (1876), 454-456. Also see: *Nature*, 15 (1877), 302.

165. ―――. "The Doctrine of Contagium Vivum and Its Application to Medicine," *Quarterly Journal of Microscopical Science*, 17 (1877), 307-329.

166. Roscoe, Sir Henry Enfield. *The Life Work of a Chemist*. Birmingham, 1889.

167. Sanderson, John Scott Burdon. [See Burdon-Sanderson, John Scott]

168. "Scientific Serials." *Nature* 14 (1876), 202.

169. Shryock, Richard Harrison. *The Development of Modern Medicine*. New York, 1947.

170. Sigerist, Henry Ernest. *The Great Doctors*. 2nd Ed. Translated by Eden and Cedar Paul. Garden City, New York, 1958.

171. Singer, Charles Joseph. *A Short History of Medicine*. Oxford, 1944.

172. Singer, Charles Joseph and Edgar Ashworth Underwood. *A Short History of Medicine.* 2nd Ed. Oxford, 1962.

173. Stanier, Roger Y., Michael Doudoroff, and Edward A. Adelberg. *The Microbial World.* 2nd Ed. Englewood Cliffs, N.J., 1963.

174. Sternberg, George Miller. *Bacteria.* New York, 1884.

175. Stich, Conrad. *Bakteriologie und Sterilisation im Apothekenbetriebe.* Berlin, 1918.

176. Sykes, George. *Disinfection and Sterilization.* 2nd Ed. London, 1965.

177. Thimann, Kenneth Vivian. *The Life of Bacteria.* 2nd Ed. New York, 1963.

178. Thomson, Wyville. "Fermentation and Putrefaction," *Nature,* 7 (1872), 61-62; 78-81.

179. Trouessart, Edouard Louis. *Microbes, Ferments, and Moulds.* New York, 1886.

180. Tyndall, John. "On Dust and Disease," *Proceedings of the Royal Institution of Great Britain,* 6 (1870), 1-14.

181. ―――. "On Dust and Smoke," *Proceedings of the Royal Institution of Great Britain,* 6 (1871), 365-376.

182. ―――. "The Optical Condition of the Atmosphere in its Bearings on Putrefaction and Infection," *Proceedings of the Royal Institution of Great Britain,* 8 (1876), 6-27.

183. ―――. "On the Optical Deportment of the Atmosphere in Reference to the Phenomena of Putrefaction and Infection," *Proceedings of the Royal Society of London,* 24 (1876), 171-183.

184. ———. "Prof. Tyndall on Germs," *Nature*, 13 (1876), 252-254; 268-270; 305.

185. ———. "The Optical Deportment of the Atmosphere in Relation to the Phenomena of Putrefaction and Infection," *Philosophical Transactions of the Royal Society of London*, 166 (1876), 27-74.

186. ———. "Note on the Deportment of Alkalized Urine," *Proceedings of the Royal Society of London*, 25 (1876), 457-458. Also see: *Nature*, 15 (1877), 302-303.

187. ———. "Preliminary Note on the Development of Organisms in Organic Infusion," *Proceedings of the Royal Society of London*, 25 (1877), 503-506.

188. ———. "On Heat as a Germicide When Discontinuously Applied," [The article consists of Tyndall's letter to Huxley.] *Proceedings of the Royal Society of London*, 25 (1877), 569-570.

189. ———. "Further Researches on the Deportment and Vital Resistance of Putrefactive and Infective Organisms, From a Physical Point of View," *Proceedings of the Royal Society of London*, 26 (1877), 228-238.

190. ———. "Further Researches on the Deportment and Vital Resistance of Putrefactive and Infective Organisms From a Physical Point of View," *Philosophical Transactions of the Royal Society of London*, 167 (1877), 149-206.

191. ———. "Note on Dr. Burdon Sanderson's Latest Views of Ferments and Germs," *Proceedings of the Royal Society of London*, 26 (1877), 353-356.

192. ———. "Observations on Hermetically-Sealed Flasks Opened on the Alps," *Proceedings of the Royal Society of London*, 26 (1877), 487-488.

193. ──────. *Fermentation and Its Bearings on the Phenomena of Disease.* London, 1877.

194. ──────. "On Schulze's Mode of Intercepting the Germinal Matter of the Air," *Proceedings of the Royal Society of London*, 27 (1878), 99-100.

195. ──────. "La Génération Spontanée," [*The Nineteenth Century*, January, 1878.] *La Revue Scientifique de la France et de l'Étranger*, 14 (1878), 1197-1210.

196. ──────. "Note on the Influence Exercised by Light on Organic Infusions," *Proceedings of the Royal Society of London*, 28 (1878), 212-213. Also see: *Nature*, 19 (1879), 210.

197. ──────. *Essays on the Floating-Matter of the Air in Relation to Putrefaction and Infection.* New York, 1888.

198. ──────. *Fragments of Science.* 2 vols. New York, 1898.

199. ──────. *New Fragments.* New York, 1898.

200. Unger, Hellmuth. *Robert Koch. (Roman eines grossen Lebens).* Berlin, 1936.

201. Vallery-Radot, René. *Louis Pasteur: His Life and Labours.* Translated by Lady Claud Hamilton. New York, 1885.

202. ──────. *The Life of Pasteur.* Translated by Mrs. R. L. Devonshire. New York, 1960.

203. Walker, Kenneth Macfarlane. *The Story of Medicine.* New York, 1954.

204. Wezel, Karl. *Robert Koch (Eine Biographische Studie)*. Berlin, 1912.

205. Williams, Henry Smith. *The Story of Nineteenth-Century Science*. New York, 1900.

206. Winslow, Charles Edward Amory. *The Conquest of Epidemic Disease. (A Chapter in the History of Ideas)*. Princeton, 1943.

207. Wolf, Abraham. *A History of Science, Technology and Philosophy in the 16th and 17th Centuries*. 2nd Ed. 2 vols. New York, 1959.

208. Wood, Laura Newbold. *Louis Pasteur*. New York, 1948.

209. Woodhead, Sir German Sims. *Bacteria and Their Products*. London, 1891.

210. Woodruff, Lorande Loss, ed. *The Development of the Sciences*. New Haven, 1923.

211. Wrench, Guy Theodore. *Lord Lister: His Life and Work*. London, 1913.

212. Wunschmann, G. "Ferdinand Julius Cohn," *Allgemeine Deutsche Biographie*, 47 (1903), 503-505.

Index

Abbott, Alexander Crever, 106, 107
biogenesis, 11, 13, 96, 105
acidic infusion, 70, 71, 90
acidic reaction, 90, 103
Ackerknecht, Erwin Heinz, 107
Adelberg, Edward A., 104, 105, 106
aerobic bacteria, 68, 73
aerobic spores, 73, 85
"aggregation of the materials,"
3
air, 84, 86
alamine racemase, 75
algae, 37, 48
alkaline infusion, 70, 71, 90
alkaline reaction, 90
ammonia, 44
anaerobic bacteria, 40, 68, 73
anaerobic spores, 73, 84
aniline blue, 17
aniline dyes, 17
aniline purple (mauve), 17
anthrax, 50, 53, 54, 56
archebiosis, 23
ascospores, 73
atmosphere, 96, 98, 99
autoclave, 81, 87

Bacillus, 39, 40, 41, 72, 79, 91
Bacillus anthracis, 11, 38, 42, 9, 57
bacillus rods, 41
Bacillus subtilis, 18, 38, 40, 41, 3, 50, 53, 54, 57, 71, 82, 90, 91

bacteria, 28, 30, 32, 33, 34, 35, 36, 38, 39, 40, 44, 47, 48, 50, 52, 54, 56, 59, 67, 70, 71, 72, 73, 75, 77, 79, 80, 90, 91, 95, 96, 99, 101, 104, 105
bacteriologists, 67
bacteriology, 16, 41, 105, 106, 110, 115
Bacterium, 29, 95
Bacterium (Streptococcus) lactis, 12, 17
Balfour, Isaac Bayley, 107
barren, 32, 82
Bary, Anton de, 107
basic fuchsin, 17
Bastian, Henry Charlton, 11, 23, 24, 26, 27, 28, 29, 30, 31, 33, 34, 36, 37, 39, 40, 41, 43, 44, 45, 46, 48, 54, 55, 56, 57, 58, 59, 60, 61, 62, 63, 64, 65, 66, 85, 90, 94, 95, 96, 98, 99, 101, 102, 103, 106, 107, 108, 109, 110
Bastian–Pasteur Paris meeting, 60-63
baths, 79, 80, 91
Berlin, 37
Bernard, Claude, 25
biogenesis, 11, 13, 95, 98, 103, 105
Biological Section of the British Association for the Advancement of Science, 95
biology of microorganisms, 91
Bochalli, Richard Max Emil Julius Theodor, 100, 110

Index

boiling, 45, 48, 79, 87, 89, 91
boiling oil, 79
Botanical Society of Edinburgh, 15
botanist, 38, 89
Boussingault, Jean Baptiste Joseph Dieudonné, 102
Breslau, 35, 37, 38, 49, 50, 97, 100
brine, 79, 80, 91
Bulloch, William, 64, 93, 94, 103, 105, 110
Bunsen burner, 27, 31, 65, 86
Bunsen, Robert Wilhelm, 36
Burdon-Sanderson, Ghetal Herschell, 95, 110
Burdon-Sanderson, John Scott, 11, 18, 23, 25, 26, 27, 28, 29, 31, 32, 33, 35, 36, 37, 38, 43, 44, 45, 46, 47, 48, 52, 53, 54, 65, 66, 69, 79, 80, 85, 89, 93, 95, 96, 98, 99, 101, 105, 106, 110, 111
Burrows, William, 104, 105, 111
butyric-acid fermentation, 41
butyric microbe, 20, 30, 40, 41, 73, 90

calcium chloride, 80
calcium dipicolinate, 75
calico, 31
Cameron, Sir Hector Clare, 111
canned foods, 81
carbolic acid, 17
carmine, 17
cast-iron box, 86
Chamberland, Charles Edouard, 63, 64, 81, 103, 111
charring, 87
Cheddar cheese, 27, 31, 40, 48
cheese, 23, 35, 40, 41, 68, 69, 95
cheese manufacture, 23, 39, 41, 89
Cheyne, Sir William Watson, 17, 93, 112
chloride of lime, 80
Clark, Paul Franklin, 112
Clifton, Charles Egolf, 94, 104, 112

closed chambers (Tyndall's), 84
Clostridium, 40, 73, 79, 91
Clostridium butyricum, 11, 38, 41, 48, 73, 89, 91
coagulation, 77, 87
Cohn, Ferdinand Julius, 11, 16, 19, 23, 30, 34, 35, 38, 39, 40, 41, 44, 45, 48, 49, 50, 52, 53, 54, 57, 73, 89, 90, 92, 93, 96, 97, 98, 99, 100, 101, 112, 113
Cohn, Pauline, 96, 100, 113
Cohn-Tyndall meeting, 50, 100
Commission of the French Academy, 60-62, 102
Conant, James Bryant, 113
conidia, 73
conidiospores, 73
consumption, 87
contamination, 30, 68, 84, 87, 90
cotton, 86
cotton-wool, 85, 86
Crace-Calvert, Frederick, 113
Creasey, C. H., 96, 115
culture dishes, 86
culture media, 80, 90

Dallinger, William Henry, 114
Dawes, Benjamin, 93, 114
dehydration, 75
De Kruif, Paul Henry, 114
denaturation, 75
de novo origin of microbes, 23, 41, 55, 90
deportment, 99, 100, 101, 102, 103, 106
desiccation, 50, 63, 90
Devonshire, Mrs. R. L., 126
Die Pflanze, 41, 97
dilution method, 12, 17
direct-flame boiling, 79
discontinuous heating, 12, 51, 54, 68, 77, 82, 84, 87, 90, 91, 101, 105
disinfection, 104
distribution, 89
Doetsch, Raymond Nicholas, 114

Index

Dolley, Charles Sumner, 105, 114
Doudoroff, Michael, 104, 105, 106
Downes, Arthur and Thomas P. Blunt, 114
dry heat, 77, 85, 86, 87
dry-oven sterilization, 77, 91
Drysdale, John James, 114
Dubos, René Jules, 102, 115
Duclaux, Émile, 102, 115
Dukes, Cuthbert Esquire, 115
Dumas, Jean Baptiste André, 61, 62
dust, 86

egg medium, 85
endospore, 48, 50, 53, 54, 57, 63, 64, 68, 71, 73, 74, 75, 79, 80, 81, 82, 83, 84, 86, 87, 89, 90, 91, 101
enzymes, 71, 74, 75, 77, 85
eosin, 17
"errors of manipulation," 48
ether, 17
evaporation, 80
Eve, Arthur Stewart, 96, 115
Ewart, J. Cossar, 115
experimental error, 90

Faraday, Michael, 36
Farmer, Laurence, 115
fermentation, 56, 92, 100, 101
Fischer, Alfred, 115
fission, 73
five minutes boiling, 11, 12, 37, 43, 45, 46, 51, 54, 55, 58, 65, 78, 79, 83, 89, 90
flasks, 86
flat sours, 81
food-preservation industry, 13, 21
Ford, William Webber, 115
Founder of Bacteriology, 41
Frankland, Edward, 25, 36, 94, 95, 115
French Academy of Medicine, 56
French Academy of Sciences, 59-62
Frobisher, Martin, 103, 104, 105, 106, 116
fungi, 24, 37, 48, 70, 72, 73, 90

Gage, Simon Henry, 116
Galdston, Iago, 116
Garnsey, Henry E. F., 107
Garrison, Fielding Hudson, 93, 116
germinate, 44, 84, 85
germinating power, 33
germ-proof container, 77
germs, 17, 18, 20, 24, 44, 47, 52, 56, 57, 58, 64, 66, 69, 70, 74, 83, 86, 90, 91, 96, 98, 99, 101, 103
germ theorists, 20, 24, 55, 56, 67
germ theory, 24, 45, 54, 56, 65, 82, 101, 105
Glasgow Address, 12, 50
glassware, 86, 87
Godlee, Mr., 79
"Golden Age of Bacteriology," 91
Gorton, David Allyn, 116
Gradle, Henry, 116
Gregory, Philip Herries, 116
Guthrie, Douglas, 116

Haagensen, Cushman Davis and Wyndham E. G. Lloyd, 116
Harms, Rudolf, 100, 116
Hartley, Walter Noel, 117
Hartnack immersion system, 16
hay, 12, 48, 49, 50, 63
hay bacillus, 73, 82, 90, 91
hay infusion, 9, 11, 12, 21, 43, 44, 47, 48, 49, 50, 51, 52, 54, 56, 65, 69, 74, 82, 90, 91
heat sensitivity, 82, 85
heat-sterilization procedures, 68, 72, 78, 82, 83, 86, 87, 89, 90, 91
Hedges, Florence, 102, 115
Henle, Friedrich Gustav Jakob, 49

Index

hermetic sealing, 46, 50, 66, 80
hermetically sealed, 80, 89, 91
heterogenesis, 90
heterogenists, 33, 41, 54, 55, 66
heterogeny, 13
Heymann, Bruno, 117
Holmes, Oliver Wendell, 117
hot-air sterilization, 87
Hueppe, Ferdinand Adolph Theophil, 117
Huizinga, Dirk, 96, 105, 117
human blood, 103
Huxley, Thomas Henry, 11, 13, 25, 51, 65, 67, 82, 83, 92, 95, 101, 117, 118

indigo, 17
infection, 96, 98, 99
infusions, 80, 82, 83, 84
"Inquirer," 37, 96, 99, 118

Jeans, William T., 118
Jodrell Laboratory, 12, 51
Johns Hopkins University, 80

Kew Gardens, 12, 51
Klein, Edward Emanuel, 104, 118
Koch, Robert, 11, 17, 41, 42, 49, 50, 53, 54, 57, 90, 91, 93, 100, 118, 119

laboratory, 87, 91
laboratory contaminants, 87
laboratory contamination, 90
laboratory glassware, 91
lactic acid bacteria, 71
lactic fermentation, 106
Lagrange, Émile, 119
Lambert, Royston, 119
Lankester, Edwin Ray, 23, 29, 30, 33, 34, 35, 36, 38, 39, 66, 68, 69, 71, 72, 89, 94, 95, 96, 119
latent period, 83
Lechevalier, Hubert A., 96, 100, 119
Leptothrix, 27, 28, 48
life-history studies, 23, 35, 37, 38, 41, 48, 49, 72, 84, 105

liquor potassae, 27, 102
Lister, Joseph, 12, 17, 20, 78, 86, 93, 105, 106, 119, 120
Liverpool Address, 11, 13, 23, 25, 95
Locy, William Albert, 120
London, 24, 61, 79
London meeting – Cohn and Tyndall, 12, 41

Marburg, 36
mauve (aniline purple), 17
meat, 81
media, 90
medicine, 100
mercury, 79
Metchnikoff, Elie, 100, 120
methyl green, 17
methyl violet, 17
microbe destruction, 85
microbes, 84, 85, 86, 89, 90
microbial specificity, 89
microbial spores, 73, 84
microbiologists, 86, 87, 90
microbiology, 21, 24, 41, 78, 81, 87, 89, 91, 104, 105
microorganisms (biology of), 83, 84, 89, 91
microphytes, 102
microscopes, 16, 31, 38, 89
Migula, Walter, 120
milk, 12, 56, 69, 78, 79, 103
milk of magnesia, 103
Milne-Edwards, Henri, 61, 62
Miquel, Pierre, 120
moist heat, 81, 85, 91
Möllers, Bernhard, 100, 120
Müller, Johannes, 37

neutral infusion (medium), 70
neutral reaction, 103
neutralization, 57, 59
Newsholme, Sir Arthur, 120
Nicolle, Jacques, 121
nitrate of soda, 80
Nordenskiöld, Erik, 121
nutritional requirements, 89

Oïdium aurantiacum, 94, 95

Index

oil, 80, 91
oil bath, 80, 84
optical condition, 98
optical deportment, 96, 98, 99
organic infusions, 96, 101, 105
organic matter, 79, 89
organic solution, 13, 21, 58, 65, 68, 72, 77, 78, 80, 89, 90, 91
organisms, 58, 100, 101, 103, 105, 106
oscillatory movements, 95
Osler, Sir William, 121
oven, 86
oven sterilization, 87
oxidation, 77
oxygen relations, 68, 73, 84, 89, 90

Papin, Denys, 81, 105
Papin digester, 32, 79, 81, 91
paraffin, 79, 80, 91
Paris, 61, 81, 87
Pasteur-Bastian experiments, 102
Pasteur Institute, 56
Pasteur, Louis, 18, 19, 25, 29, 34, 39, 40, 54, 55, 56, 57, 59, 60, 61, 62, 63, 64, 66, 69, 70, 73, 80, 90, 93, 97, 99, 101, 102, 103, 121, 122
Pasteur, Louis and Jules François Joubert, 122
pathogen, 50, 54
pathogenicity, 89
pathology, 106
Payen, Anselme, 94, 95, 122
pea infusions, 34, 39, 49
Penicillium, 70, 95
peptone, 75
peptone water, 71
Perkin, William Henry, 17, 122
pH reaction, 68, 69, 70, 71, 73, 75, 90, 103
pipettes, 86
Plant Physiology Institute, 38
plates, 86
platinum loop, 85
pleomorphism, 20

pleomorphy, 16
Pode, C. C., 29, 33, 35, 96, 122
"polar emergence," 73
potash, 57
potash solution, 56, 57, 58, 60
potassium carbonate, 27, 102
pre-existing germs, 89
pregnant, 32, 33
protein, 71, 74, 77, 85
pure culture, 12, 20, 89
putrefaction, 92, 96, 98, 99

rabies, 56
Rayleigh, John William Strutt (Lord), 122
rennet infusion, 40
researches, 45, 100, 101, 103, 106
Roberts, William, 11, 17, 18, 19, 30, 31, 34, 38, 44, 45, 46, 47, 48, 49, 50, 52, 53, 54, 56, 57, 58, 59, 65, 67, 68, 69, 72, 78, 79, 80, 85, 89, 90, 93, 95, 96, 98, 102, 103, 104, 105, 106, 123
Roscoe, Sir Henry Enfield, 123
Royal Institution, 12, 36, 50, 51, 68, 98
Royal Society of Edinburgh, 20
Royal Society of London, 12, 30, 44, 51, 57, 59, 82, 96

safranin, 17
Sanderson, John Scott Burdon (see Burdon-Sanderson, John Scott), 123
"Scientific Serials," 97, 123
scientists, 89, 92
serum medium, 85
sheet iron, 86
Smith, Erwin F., 102, 115
Shryock, Richard Harrison, 123
Sigerist, Henry Ernest, 123
Singer, Charles Joseph, 123, 124
solid particles, 90
Solotorovsky, Morris, 96, 100
specific gravity, 31, 43, 45, 60, 79
spirit lamp, 84

Index

spontaneous generation, 13, 53, 54, 56, 59, 60, 64, 89, 90, 91, 101, 102
sporangiospores, 73
spore formation, 40, 48, 53, 54, 67
spore germination, 73
sputtering, 79
staining procedures, 17, 89
Stanier, Roger Y., 104, 105, 106, 124
steam, 87
steam under pressure, 81
sterilization, 67, 81, 84, 86, 90, 91, 104
Sternberg, George Miller, 77, 80, 104, 105, 124
Stich, Conrad, 124
stove, 86
superheated steam, 81
surgery, 100
Sykes, George, 81, 94, 103, 104, 105, 106, 124

test tubes, 86
thermal death point, 33, 35, 66, 67, 68
thermal death time, 66, 67, 68
thermal resistance, 90, 91
thermometer, 86
thermophilic bacteria, 75
Thimann, Kenneth Vivian, 93, 104, 124
Thomson, Wyville, 15, 16, 92, 124
time, 91
torulae (yeast-like fungi), 34, 72, 95
Trouessart, Edouard Louis, 124
turnip-cheese episode, 23-42, 71, 72, 89
turnip-cheese experiments, 96
turnip-cheese infusion, 21, 23, 24, 25, 29, 31, 32, 33, 36, 38, 39, 41, 47, 48, 49, 52, 65, 66, 68, 69, 73, 79, 91
turnip infusion, 29, 95
turnips, 35, 48

Tyndall, John, 11, 18, 23, 27, 36, 37, 38, 41, 44, 45, 46, 47, 48, 49, 50, 51, 52, 53, 54, 55, 57, 58, 59, 62, 63, 65, 66, 67, 68, 69, 70, 74, 78, 82, 83, 84, 89, 90, 91, 93, 95, 96, 98, 99, 100, 101, 102, 103, 104, 105, 106, 124, 125, 126
Tyndall, Lady Claud Hamilton, 101, 103
Tyndallization, 51, 54, 77, 82, 84, 85, 87

Underwood, Edgar Ashworth, 124
Unger, Hellmuth, 126
United States Pharmacopeia, 86
University College, London, 24, 25, 29, 79
urine, 21, 55, 56, 57, 58, 59, 60, 64, 65, 69, 70, 90, 101, 102

Vallery-Radot, René, 63, 101, 102, 103, 126
Van Tieghem, Philippe Edouard Léon, 62, 102
vegetative cell, 73, 74, 82, 83
vinegar, 103
vital persistence, 99, 100, 101, 103, 106

Walker, Kenneth Macfarlane, 126
water, 84
water-bath technique, 79
Weigert, Carl, 17
Weisnegg engineering company, 81
Wezel, Karl, 127
Williams, Henry Smith, 127
Winslow, Charles Edward Amory, 127
Wolf, Abraham, 105, 127
Wood, Laura Newbold, 127
Woodhead, Sir German Sims, 127
Woodruff, Lorande Loss, 127
wool plugs, 81
Wrench, Guy Theodore, 127
Wunschmann, G., 97, 127

Giambattista della Porta

**On Spontaneous Generation
(1589)**

being

Book II: "Of the Generation of Animals"
(Chapters I — IV)
and
Book III: "Of the Production of New Plants"
(Chapter I)

of
Giambattista della Porta
Magiae naturalis libri XX
(Naples, 1589)

Anonymous English translation
Natural Magick
(London, 1658)

The Proeme

Having wandred beyond my bounds, in the consideration of Causes and their Actions; which I thought fit to make the Subject of my first book: it will be time to speak of those Operations, which we have often promised, that we may not too long keep off from them those ingenious men that are very desirous to know them. Since that we have said, That Natural Magick is the top, and the compleat faculty or Natural Science, in handling it, we will conclude within the compass of this Volume whatsoever is High, Noble, Choice, and Notable, that is discovered in the large field of Natural History. But that we may perform this, I shall reduce all those Secrets into their proper places; and that nothing may be thrust out of its own rank, I shall follow the order of Sciences. And I shall first divide them into Natural and Mathematical Sciences; and I shall begin with the Natural; for I hold that most convenient, that all may arise from those things that are simple, and not so laborious, to Mathematical Sciences. I shall from Animals first proceed to Plants, and so by steps to Minerals, and other works of Nature. I shall briefly describe Fountains, also whence flow Springs; and I shall annex thereto the Reasons, and the Causes: that Industrious men made acquainted with this, may find out more of themselves. And because there are two generations of Animals and Plants, one of themselves, the other by copulation: I shall first speak of such as are bred without copulation; and next, of such as proceed from copulation one with another, that we may produce new living Creatures, such as the former ages never saw. I shall begin therefore with Putrefaction, because that is the principle to produce new Creatures; not onely from the variety of Simples, but of mixed Bodies. I thought fit to leave none out, though they be of small account, since there is nothing in Nature, appear it never so small, wherein there is not something to be admired.

Chapter I
The first Chapter treateth of Putrefaction, and of a strange manner of producing living Creatures.

Before we come to shew that new living Creatures are generated of Putrefaction, it is meet to rehearse the opinions of antient Philosophers concerning that matter Whereof though we have spoken elsewhere, in the description of Plants, yet for the Readers ease, we will here rehearse some of them, to shew that not onely imperfect, but perfect living Creatures too, are generated of Putrefaction. *Porphyry* thought that Living creatures were begotten of the bowels of the Earth soaked in water, and quickened by the heat of the Sun. Of the same mind were *Archelaus* the Athenian, *Anaxagoras Clazomenius*, and *Euripides* his Scolar. *Cleodemus*, and after him *Theophrastus*, thought that they came of putrified water mixt with earth; and the colder and fouler the water was, the unfitter it was for their generation. *Diodorus*, and many other good Philosophers hold, that all living Creatures did arise of putrefaction. For whereas in the beginning of the world, the Heavens, and Earth, and Elements were settled in their natural places, the earth being left slimy and soft in many places, and then dried and stricked with the heat of the Sun, brought forth certain tumors and swellings in the surface and uppermost parts: in these tumors were contained and cherished many putrefactions and rotten clods, covered over with certain small skins; this putrified stuff, being moistened with dew by night, and the Sun heating it by day, after a certain season became ripe; and the skins being broken, thence issued all kinds of living Creatures; whereof, they that had quickest heat, became birds; the earthy ones became creeping beasts; the waterish ones became fishes in the Sea; and they which were a mean, as it were, betwixt all these, became walking-creatures. But the heat of the Sun still working upon the earth, hindered it from begetting and bringing forth any more such creatures; but then, the creatures before generated coupled together, and brought forth others like themselves. *Avicenna*, in that work of his which he made of deluges and flouds; holds, that after the great flouds that drowned the Earth, there was no mans seed; but then, man, and all living Creatures else, were generated of rotten carcases, only by the vertue of the Sun: and therefore he supposeth, that the womb, and such needful places framed by nature, for the better fashioning of the infant, are not needfull to the procreation of man. He proves his assertion by this, that mice, which arise of putrefaction, do couple together, and beget store of young; yea, and serpents are generated chiefly of womans hair. And in his book of living Creatures, he tels of a friend of his, that brought forth Scorpions after a strange manner, and those did beget other Scorpions, not imperfect, or unlike to themselves, but such as did also procreate others. *Averroes* held, that the stars were sufficient to generate imperfect creatures; as mice, bats, moules, and such like, but not to generate Men, or Lions. And daily experience teacheth us, that many living creatures come of the putrified matter of the earth. And the Ancients supposing all things to be produced out of the earth, called it the mother of all; and the Greeks called it Dimitera. *Ovid* hath very elegantly set down this generation of putrefaction, under the fable of *Pytho*; that the earth brought forth of its own accord, many living creatures of divers forms, the heat of the Sun enliving those moistures that lay in the tumors of the earth, like fertile seeds in the belly of their mother; for heat and moisture being tempered together, causeth generation. So then, after the deluge, the earth being now moist, the Sun working upon it, divers kinds of creatures were brought forth, some like the former, and some of a new shape.

Chapter II
Of certain earthly Creatures, which are generated of putrefaction.

Plants and living Creatures agree both in this, that some of them are generated of seed, and some of them Nature brings forth of her own accord, without any seed of the same kind; some out of putrified earth and plants, as those Creatures that are divided between the head and the belly; some out of the dew that lies upon leaves, as Canker-worms; some out of the mud, as shel-creatures; and some out of living Creatures themselves, and the excrements of their parts, as lice. We will onely rehearse some which the Ancients have set down, that so we may also learn how to procreate new creatures. And first, let us see, how

Mice are generated of putrefaction.
Diodorus saith, that neer to the City Thebais in Egypt, when Nilus overflowing is past, the Sun heating the wet ground, the chaps of the earth send forth great store of mice in many places; which astonisheth men to see, that the fore-part of the mice should live and be moved, whereas their hinder parts are not yet shapen. *Pliny* saith, that after the swaging of Nilus, there are found little mice begun to be made of earth and water, their fore-parts living, and their hinder parts being nothing but earth. *Aelianus* saith, that a little rain in Egypt, engenders many mice, which being scattered everywhere in their fields, eat down their corn, and devour it: And so it is in Pontus; but by their prayers to God, they are consumed. *Macrobius* and *Avicenna* say, that the mice so generated, do encrease exceedingly by coupling together. *Aristotle* found out, that a kind of field-mice encreased wonderfully; so that in some places they did suddenly eat up whole fields of corn: insomuch that many Husband-men appointing to reap their corn on the morrow, when they came with their reapers, found all their corn wasted. And as

these mice are generated suddenly, so they are suddenly consumed, in a few dayes; the reason whereof cannot be so well assigned. *Pliny* could not find how it should be; for neither could they be found dead in the fields, neither alive within the earth in the winter time. *Diodorus* and *Aelianus* write, That these field-mice have driven many people of Italy out of their own Countrey: they destroyed Cosas, a City of Hetruria: many came to Troas, and thence drove the inhabitants. *Theophrastus* and *Varro* write, That mice also made the inhabitants of the Island Gyarus to forsake their Country; and the like is reported of Heraclea in Pontus, and of other places. Likewise also

Frogs are wonderfully generated of rotten dust and rain;

for a Summer showre lighting upon the putrified sands of the shore, and dust of high-wayes, engenders frogs. *Aelianus*, going from Naples in Italy, to Puteoli, saw certain frogs, that their fore-parts moved and went upon two feet, while yet their hinder parts were unfashioned, and drawn after like a clot of dirt: and *Ovid* saith, one part lives, the other is earth still: and again, mud engenders frogs that sometimes lack feet. The generation of them is so easie, and sudden, that some write it hath rained frogs; as if they were gendred in the Air. *Phylarchus* in *Athenaus* writes so; and *Heraclides Lembus* writes, that it rained frogs about Dardany and Poeonia, so plentifully, that the very wayes and houses were full of them: and therefore the inhabitants, though for a few daies at the first they endured it, killing the frogs, and shutting up their houses, yet afterward when they saw it was to no purpose, but they could neither use water, nor boil meat, but frogs would be in it, nor so much as tread upon the ground for them, they quite forsook their countries, as *Diodorus* and *Eustathius* write. The people Autharidae in Thespratia, were driven out of their Country, by certain imperfect frogs that fell from heaven. But it is a strange thing that

Red Toads are generated of dirt, and of womens flowers.

In Dariene, a Province of the new world, the air is most unwholesome, the place being muddy and full of stinking marishes; nay, the village is it self a marish, where Toads are presently gendred of the drops wherewith they water their houses, as *Peter Martyr* writes. A Toad is likewise generated of a duck that hath lyen rotting under the mud, as the verse shews which is ascribed to the duck; When I am rotten in the earth, I bring forth Toads: happily because they and I both, are moist and foul creatures. Neither is it hard to generate Toades of womens putrified flowers; for women do breed this kind of cattel, together with their children, as *Celius Aurelianus* and *Platearius* call them, frogs, toads, lyzards, and such like: and the women of Salerium, in times past, were wont to use the juice of Parsley and Leeks, at the beginning of their conception, and especially about the time of their quickening, thereby to destroy this kind of vermin with them. A certain woman lately married, being in all mens judgement great with child, brought forth in stead of a child, four Creatures like to frogs, and after had her perfect health. But this was a kind of a Moon-calf. *Paracelsus* said, that if you cut a serpent in pieces, and hide him in a vessel of glasse, under the mud, there will be gendred many worms, which being nourished by the mud, will grow every one as big as a Serpent; so that of one serpent may be an hundred generated: and the like he holds of other creatures. I will not gainsay it, but only thus, that they do not gender the same serpents. And so, he saith, you may make them of a womans flowers; and so, he saith, you may generate a Basilisk, that all shall die which look upon him: but this is a stark lie. It is evident also, that

> *Serpents may be generated of mans marrow,*
> *of the hairs of a menstruous woman,*
> *and of a horse-tail, or mane.*

We read, that in Hungary, by the River Theisa, Serpents and Lyzards did breed in mens bodies, so that three thousand men died of it. *Pliny* writes, that about the beginning of the wars against the Marsi, a maid-servant brought forth a serpent. *Avicenna* in his book of deluges, writes, that serpents are gendred of womens hairs especially, because they are naturally moister and longer then mens. We have experienced also, that the hairs of a horses mane laid in the waters, will become serpents: and our friends have tried the same. No man denies but that serpents are easily gendred of mans flesh, especially of his marrow. *Aelianus* saith, that a dead mans back-marrow being putrified, becomes a serpent: and so of the meekest living Creature arises the most savage: and that evil mens back-bones do breed such monsters after death; *Ovid* shews, that many hold it for a truth. *Pliny* received it of many reports, that Snakes gendred of the marrow of mens backs. Writers also shew,

> *How a Scorpion may be generated of Basil.*

Florentinus the Grecian saith, That Basil chewed and laid in the Sun, will engender serpents. *Pliny* addeth; that if you rub it, and cover it with a stone, it will become a Scorpion; and if you chew it, and lay it in the Sun, it will bring forth worms. And some say, that if you stamp a handful of Basil, together with ten Crabs or Crevises, all the Scorpions thereabouts will come unto it. *Avicenna* tells of a strange kind of producing a Scorpion; but *Galen* denies it to be true. But the body of a Crab-fish is strangely turned into a Scorpion: *Pliny* saith, that while the Sun is in the sign Cancer, if the bodies of those fishes lie dead upon the Land, they wil be turned into Scorpions. *Ovid* saith, if you take off the Crabs arms, and hide the rest in the ground, it will be a Scorpion. There is also a

Creature that lives but one day, bred in vineger;

as *Aelianus* writes; and it is called Ephemerus, because it lives but one day: it is gendred of the dregs of sowre wine; and as soon as the vessel is open, that it comes into the light, presently it dies. The River *Hippanis*, about the solstitial daies, yields certain little husks, whence issue forth certain four-footed birds, which live and flie about till noon, but pine away as the Sun draws downward, and die at the Sun-setting; and because they live but one day, they are called Hemerobion, a daies-bird. So the

Pyrigones be generated in the fire;

Certain little flying beasts, so called, because they live and are nourished in the fire; and yet they flie up and down in the Air. This is strange; but that is more strange, that as soon as ever they come out of the fire, into any cold air, presently they die. Likewise the

Salamander is gendred of the water;

for the Salamander it self genders nothing, neither is there any male or female amongst them, nor yet amongst Eeels, nor any kind else; which doth not generate of themselves either egge or young, as *Pliny* noteth. But now we will speak of a most excellent generation, namely, how

Bees are generated of an Ox.

Aelianus writes, That Oxen are commodious many wayes; amongst the rest, this is one excellent commodity, that being dead, there may be generated of them a very profitable kind of Creatuers, namely Bees. *Ovid* saith it, that as all putrified bodies are turned into some small living Creatuers, so Oxen putrified do generate Bees. *Florentinus* the Grecian saith, that *Jubas* King of Africa, taught how to make Bees in a wooden Ark. *Democritus* and *Varro* shew a cruel manner of making Bees in a house: but it is a very ready way. Chuse a house ten cubits high, and ten cubits broad, square every way: but let there be but one entrance into it, and four windows, on each side one. Put in this room an Ox, about two or three years old; let him be fat and fleshy: then set to him a company of lusty fellows, to beat him so cruelly, that they kill him with their cudgels, and break his bones withal: but they must take great heed that they draw no blood of him, neither must they strike him too fiercely at the first: After this, stop up all the passages of the Ox, his nostrils, eyes, mouth, and necessary places of evacuation, with fine linen clouts besmeared with pitch: Then cast a great deal of honey under him, being laid with his face upwards, and let them all go forth, and daube up the door and the windows with thick lome, so that no wind, nor Air can get in. Three weeks after, open the room, and let the light and the Air come in, except there where the wind would blow in too violently. And when you see that the matter is through cold, and hath taken air enough, then shut up the door and windows as before. About eleven daies after, open it again, and you shall find the room full of Bees clotted together, and nothing of the Ox remaining, beside the horns, the bones and the hair. They say that the Kings of the companies are generated of the brain, the other of the flesh, but the chief Kings of all, of the marrow; yet those that come of the brain, are most of them greater, handsomer, and better-coloured then the rest. When you open the room first, you shall find the flesh turned into small, white, and unperfect creatures, all of the same shape, but as yet only growing, and not moving. Afterward, at the second opening, you may see their wings grown, the right colour of Bees in them, and how they sit about their Kings, and flutter about, especially toward the windows, where they would enjoy their desired light. But it is best to let them light by the windows every other day. This same experiment, *Virgil* hath very elegantly set down in the same manner. Now as the best kind of Bees are generated of a young Ox, so a more base kind of them is brought forth of the dead flesh of baser creatures; *Aelianus* saith,

That Waspes are generated of an Horse;

when his carcase is putrified, the marrow of him brings forth Waspes; a swift kind of fowl, from a swift kind of beast. *Ovid* saith, that Hornets are thence generated; and *Isodore* derives *crabronem á cabo, id est caballo,* a hornet of a horse, because they are brought forth of horses. *Pliny* and *Virgil* say, that waspes and hornets both, are generated of the flesh of dead horses. In like manner

Drones come of Mules,

as *Isodore* affirmeth: and the Drone is called *Fucus quasi Fagos,* because he eats that which he never laboured for. But others hold that Locusts, and not Drones, are generated of Mules flesh. So also, of the basest beast cometh the basest fowl:

The Beetle is generated of the Ass,

as *Pliny* writes. *Isodore* saith, they come of swift dogs: *Aelianus* saith, they have no female, but lay their seed in a clot of earth for 28 dayes, and then bring forth young out of it.

Chapter III
Of certain Birds, which are generated of the Putrefaction of Plants

Olaus Magnus, in the description of the North-countries of Europe, reports, that about Scotland, there be certain birds generated of the fruit of a Tree. *Munster* saith, there be certain Trees which bring forth a fruit covered over with leaves; which, if it fall into the water under it, at the right season, it lives, and becomes a quick bird, which is called *Avis arborea*. Neither is this any new tale; for the antient Cosmographers, especially *Saxo Grammaticus* mentions the same Tree. Late Writers report, That not onely in Scotland, but in the River of Thames also by London, there is a kind of Shel-fish in a two-leaved shell, that hath a foot full of plaits and wrinkles: these fish are little, round, and outwardly white, smooth and brittle shelled, like an Almond shell; inwardly they are great bellied, bred as it were of moss and mud: they commonly stick on the keel of some old Ship, where they hang together like Mushrome-stalks, as if they were thereby nourished. Some say, they come of worms, some of the boughs and branches of Trees which fall into the Sea; if any of these be cast upon shore, they die; but they which are swallowed still into the Sea, live, and get out of their shell, and grow to be ducks or such like birds. *Gesner* saith, that in the Islands Hebrides, the same

Birds are generated of putrified wood.

If you cast wood into the Sea, first after a while there will certain worms breed in it, which by little and little become like ducks, in the head, feet, wings and feathers; and at length grow to be as big as Geese: and when they are come to their full growth, they flie about in the Air, as other birds do. As soon as the wood begins first to be putrified, there appears a great many wormes, some unshapen, others being in some parts perfect, some having feathers, and some none. *Paracelsus* saith; As the yelk and white of an egge, becomes a chick by the heat of an Hen; so a bird burnt to ashes, and shut up in a vessel of glass, and so laid under the mixen, will become a slimy humour; and then, if it be laid under a Hen, is enlived by her heat, and restored to her self like a Phoenix. *Ficinus* reporteth, and he had it out of *Albertus*, That there is a certain bird, much like a Black-bird, which is generated of the putrefaction of Sage; which receives her life and quickning from the general life of the whole world.

Chapter IV
Of Certain fishes which are generated of putrefaction

Having first spoken of earthly Creatures, and then of Fowles; now we will speak of Fishes so generated. And first how

Eeles are generated.

Amongst them there is neither male or female, nor egges, nor any copulation; neither was there ever seen in any of them, any passage fit to be a womb. They have bred oft-times in certain muddy pools, even after all the water and mud hath been gone; only by rain-water: neither indeed do they ever breed without rain, though they have never so much water otherwise; for it is the rain, both that begets and nourishes them, as *Aristotle* writes. They are also generated of putrified things. Experience hath proved, that a dead horse thrown into a standing pool, hath brought forth great store of Eeles; and the like hath been done by the carcases of other creatures. *Aristotle* saith, they are generated of the garbage of the earth, which he saith, ariseth in the Sea, in Rivers, and in pools, by reason chiefly of putrefaction; but it arises in the Sea by reason of reeds; in Pools and Rivers, it arises by the banks-side, for there the heat is more forcible to cause putrefaction. And a friend of mine filled certain wooden vessels with water, and Reeds, and some other water-herbs, and set them in the open Air, having first covered them with a weighty stone, and so in short time generated Eeles. Such is the generation of

Groundlings out of fome and froth,

which fish the Greeks call Aphya, because rain breeds it. Many of them breed of the fome that rises out of the sandy chanel, that still goes and comes at all times, till at last it is dissolved; so that this kind of fish breeds all times of the year, in shadowy and warm places, when the soyl is heated; as in Attica, neer to Salamnia, and in Marathon, where *Themistocles* got his famous victory. In some places, this fish breeds of fome by the help of the rain; and swims on the top of the water in the fome, as you see little wormes creep on the top of mud. *Athenaeus* saith, This fish is consecrated to Venus, because she also comes of the froth of the Sea, whence she is called Aphrodites. *Aelianus* saith, These fishes neither do beget, nor are begotten, but only come of mud: for when dirt is clotted together in the Sea, it waxes very black and slimy, and then receives heat and life after a wonderful manner, and so is changed into very many living Creatures, and namely into Groundlings. When the waves are too boistrous for him, he hides himself in the clift of some rock; neither doth he need any food. And *Oppianus* makes the very same description of them, and of their generation. There is a kind of these fishes, called a Mullet-Groundling, which is generated of mud and of sand, as hath been tried in many marish places, amongst the rest in Gindus; where in the Dog-daies, the Lakes being dried up, so that the mud was hard; as soon as ever they began to be full of rain-water again, were generated little fishes, a kind of Mullets, about the bigness of little Cackrels, which had neither seed nor egge in them. And in some parts of Asia, at the mouth of the Rivers into the Sea, some of a bigger size are generated. And as the Mullet-groundling comes of mud, or of a sandy lome, as *Aristotle* writes; so it is to be thought, that the Cackrel-groundling comes thereof also. It seems too, that

A Carpe is generated of putrefaction,

Especially of the putrified mud of sweet water: for it is experienced, that in certain Lakes, compassed about with Hills, where there is no Well, nor River, to moisten it, but only the rain, after some few showers, there hath been great store of fish, especially Carps: but there are some of this kind generated by copulation. There are also in certain particular Lakes, particular kinds of fishes, as in the Lemane, and the Benacian Lakes, there be divers kind of Carpes, and other such fishes. Likewise there are certain

Earthly fishes generated of putrefaction.

Pliny reports, that in Paphlagonia, they dig out of deep ditches, certain earthly fishes very good to be eaten; and it is so in places where there is no standing water; and he wonders that they should be generated without copulation;: but surely it is by vertue of some moisture, which he ascribes to the Wells, because in some of them fishes are found. Likewise

Shel-fish are generated of the frothy mud,

or else meerly of the salt-water; for they have neither seed, nor male, nor female; the hardnesse and closenesse of their shells, hindering all things from touching or rubbing their inward parts, which might be fit for generation. *Aristotle* saith, they breed all of themselves; which appears by this, that oft-times they breed in Ships, of a forthy [frothy] mud putrified: and in many places, where no such thing was before, many shel fishes have bred, when once the place waxed muddy, for lack of moisture. And that these fishes emit no seed or generative matter, it appears, because that when the men of Chios had brought out of Lesbos many Oysters, and cast them into Lakes neer the Sea, there were found no more then were cast in; onely they were somewhat greater. So then Oysters are generated in the Sea, in Rivers and in Lakes, and therefore are called Limnostrea, because they breed in muddy places. *Oppianus* writes also, that they have neither male nor female, but are generated of themselves and their own accord, without the help of any copulation. So the fish called *Ortica*, and the Purple, and Muscles, and Scallops, and Perwinkles, and Limpins, and all Shel-fish are generated of mud: for they cannot couple together, but live only as plants live. And look how the mud differs, so doth it bring forth different kinds of fishes: durty mud genders Oysters, sandy mud Perwinkles, the mud in the Rocks breedeth Holoturia, Lepades, and such-like. Limpins, as experience hath shewed, have bred of rotten hedges made to fish by; and as soon as the hedges were gone, there have been found no more Limpins.

Chapter I
How new kinds of Plants may be generated of putrefaction

As we have shewed before, that new kinds of Living Creatures may be generated of putrefaction; so, to proceed in the same order as we have begun, we will now shew that new kinds of Plants may grow up of their own accord, without any help of seed or such like. The Antients questionless were of opinion, that divers plants were generated of the earth and water mixt together; and that particular places did yield certain particular plants. We rehearsed the opinion of *Diogenes* before, who held that plants are generated of water putrified in it self, and a little earth tempered therewith. *Theophrastus* held, that the rain causeth much putrefaction and alteration in the earth, and thereby plants may be nourished, the Sun working upon it with his heating, and with his drying operation. They write also, that the ground when it is stirred, brings forth such kinds of Plants alwaies, as are usuall in the same place. In the Isle Creta, the ground is of that nature, that if it be stirred anywhere, and no other thing sown or planted in it, it will of it self bring forth a Cypresse tree: and their tilled lands, those that are somewhat moist, when they lie fallow, bring forth thistles. So the herb Laser in Africa, is generated of a kind of pitchy or clammy rain and thick dirt; and the herb will shew it self out of the earth presently after the rain is fallen. *Pliny* said, that the waters which fall from above, are the cause of every thing that grows upon the earth, nature shewing therein her admirable work and power: and many such things they report, which we have spoken of in the books of the knowledge of Plants. And I my self have oft-times by experience proved, that ground digged out from under the lowest foundations of certain houses, and the bottom of some pits, and laid open in some small vessel to the force of the Sun, hath brought forth divers kinds of Plants. And whereas I had oftentimes, partly for my own pleasure, and partly to search into the works of Nature, sought out and gathered together earth of divers kinds, I laid them abroad in the Sun, and watered them often with a little sprinkling, and found thereby, that a fine light earth would bring forth herbs that had slight stalkes like a rush, and leaves full of fine little ragges; and likewise that a rough and stiff earth full of holes, would bring forth a slight herbe, hard as wood, and full of crevises. In like manner, if I took of the earth that had been digged out of the thick woods, or out of moist places, or out of the holes that are in hollow stones, it would bring forth herbs that had smooth blewish stalkes, and leaves full of juice and substance, such as Peny-wort, Purslane, Senegreek, and Stone-croppe. We made trial also of some kinds of earth that had been farre fetcht, such as they had used for the ballast of their Shippes; and we found such herbs generated thereof, as we knew not what they were. Nay further also, even out of very roots and barks of

Trees, and rotten seeds, powned and buried, and there macecrated with water, we have brought forth in a manner the very same herbs; as out of an Oken root, the herb Polypody, and Oak-fern, and Splenewort, or at least such herbs as did resemble those, both in making and in properties. What should I here rehearse, how many kinds of toad-stools and puffs we have produced; yea, of every several mixture of putrified things, so many several kinds have been generated. All which I would here have set down, if I could have reduced them into any method; or else if such plants had been produced, as I intended: but those came that were never sought for. But happily I shall hereafter, if God will, write of these things, for the delight, and speculation, and profit of the more curious sort: which I have neither time nor leisure now to mention, seeing this work is ruffled up in haste. But let us see

How Toad-stools may be generated

Dioscorides, and others have written, That the bark of a white Poplar-Tree, and of a black, being cut into small pieces, and sowed in dunged lands or furrows, will at all times of the year bring forth mushromes or toad-stools that are good to be eaten. And in another place he saith, that they are more particularly generated in those places, where there lies some old rusty iron, or some rotten cloth: but such as grow neer to a Serpents hole, or any noisome Plants, are very hurtful. But *Tarentinus* speaks of this matter more precisely. If, saith he, you cut the stock of a black Poplar peece-meal into the earth, and pour upon it some leaven that hath been steeped in water, there will soon grow up some Poplar toad-stools. He addeth further; If an up-land or hilly field that hath in it much stubble and many stalks of corn, be set on fire at such time as there is rain brewing in the clouds, then the rain falling, will cause many toad-stools there to spring up of their own accord: but if, after the field is thus set on fire, happily the rain which the clouds before threatned doth not fall; then, if you take a thin linnen cloth, and let the water drop through by little and little like rain, upon some part of the field where the fire hath been, there will grow up toad-stools, but not so good as otherwise they would be, if they had been nourished with a showre of rain. Next we will shew

How Sperage may be generated.

Dydimus writes, That if any man would have good store of Sperage to grow, he must take the horns of wilde Rams, and beat them into very small powder, and sow them in eared ground, and water it, and he shall have his intent. There is one that reports a more strange matter; that if you take whole Rams horns not powned into small pieces, but only cut a little, and make a hole in them, and so set them, they will bring forth Sperage. *Pliny* is of *Didymus* opinion, that if the horns be powned and digged into the earth, they will yield Sperage; though

Dioscorides thinks it to be impossible. And though I have made often trial hereof, but could not find it so to be, yet my friends have told me of their own experience, that the same tender seed that is contained within the Rams horn, hath produced Sperage. The same my friends also have reported

That Ivy doth grow out of the Harts horn;
and *Aristotle* writes of an Husband-man that found such an experiment; though for my own part I never tried it. But *Theophrastus* writes, that there was Ivy found growing in the Harts horn; whereas it is impossible to think how any Ivy seed could get in there: and whereas some alledge, that the Hart might have rubbed his horn against some Ivy roots, and so some part of the horn being soft and ready to putrifie, did receive into it some part of the root, and by this means it might there grow; this supposal carries no shew of probability or credit with it. But if these things be true, as I can say or see nothing to the contrary, then surely no man will deny but that divers kinds of plants may be generated of divers kinds of living Creatures horns. In like manner, may plants be generated of the putrified barks and boughs of old Trees: for so is

Polypody, and the herb Hyphear generated;
for both these, and divers other plants also, do grow up in Firre-trees, and Pine-trees, and such other; for in many Trees, neer to the bark, there is a certain flegmatick or moist humour, that is wont to putrifie; which, when it abounds too much within, breaks forth into the outward shew of the boughs and the stock of the Tree; and there it meets with the putrified humour of the bark; and the heat of the Sun working upon it there, quickly turns it into such kinds of herbs.

†

Monsters born from the Deluge

Wood Engravings in the
Chronique de Nuremberg
1493